# WEIGHING EVIDENCE IN LANGUAGE AND LITERATURE

## A statistical approach

BARRON BRAINERD

In recent years, there has been a tremendous development in the area of quantitative and statistical analysis of linguistic and literary data, generated, no doubt, by extensive advances in computer technology and their relatively easy availability to scholars. However, except for a few rather specialized examples, there has been no truly introductory text in statistics and quantitative analysis devoted to the needs of language scholars. This work was written especially to fill the gap. It introduces a mathematically naïve reader to those statistical tools which are applicable in modern quantitative text and language analysis, and does this in terms of simple examples dealing exclusively with language and literature. Exercises are included throughout.

BARRON BRAINERD is professor of mathematics and an active member of the Centre for Linguistic Studies at the University of Toronto. He is co-author of several books on mathematics, has contributed to volumes on mathematical and statistical linguistics, and has published articles in international journals.

T0260450

# MATHEMATICAL EXPOSITIONS

*Editorial Board*

H.S.M. COXETER, G.F.D. DUFF, D.A.S. FRASER,
G. DE B. ROBINSON (Secretary), P.G. ROONEY

*Volumes Published*

MATHEMATICAL EXPOSITIONS No. 19

# WEIGHING EVIDENCE IN LANGUAGE AND LITERATURE:
## A statistical approach

**BARRON BRAINERD**

UNIVERSITY OF TORONTO PRESS

© University of Toronto Press 1974
Toronto and Buffalo
Reprinted in paperback 2014
ISBN 978-0-8020-1874-8 (cloth)
ISBN 978-1-4426-5222-4 (paper)
CN ISSN 0076–5333
LC 70–190342
AMS 1970 Subject Classifications 62 and 94.50

# Contents

# Preface

In recent years, a large number of studies devoted to the quantitative and statistical analysis of linguistic and literary data have appeared, as the reader will discover if he consults the annotated bibliography of Bailey and Doležel (1968) and the collections of articles edited by Doležel and Bailey (1969), Garvin (1964), and Leed (1966). By and large this awakening of interest in quantitative and statistical methods among humanists has not been accompanied by any particularly deep understanding of the methodology in these areas. The tendency has been, for the most part, for humanists to apply 'recipes' to their data without particular consideration of their suitability. There are, nevertheless, a number of statistically rather sophisticated studies in the area – for example Dyen et al. (1967), Elderton (1949), Yule (1944), and Mosteller and Wallace (1964) – although, with the possible exception of Yule (1944), these are of little use to the statistically naïve humanist.

A few statistical textbooks, and near textbooks, directed mainly toward students of literary style have been published – for example, to name some of the better known, Somers (1962), Guiraud (1959), Muller (1968), and a series of works by Gustav Herdan, the last of which was published in 1966. In addition, there are many textbooks of statistics without any particular language bias. However, most, if not all, of the language-oriented texts are incomplete or superficial or both in their coverage of statistical methodology, whereas the classical statistical texts appear to have failed to motivate the humanistic reader to achieve a sufficiently intuitive grasp of the probabilistic underpinnings of statistics to use the methods with power and confidence. The writing of the present work was undertaken with a view to filling the gap, by providing an introduction to the statistical approach to data for use in the study of language and style. It is hoped that the volume will make the innocent user of statistics a little less innocent.

The treatment is entirely introductory, requiring only a modest knowledge of algebra and simple set theory at the high school level. A remedial chapter, Mathematical Preliminaries, is provided for those who feel that they might not meet the mathematical prerequisites. I have tried, too, to keep the mathematics in the volume to a minimum. Where a mathematical development has been included, it is thought to be essential for the understanding of the topic or of particular interest for those with a knowledge of the calculus. In such instances the reader without calculus may omit the mathematical development and rejoin the discussion when the result is reached.

The book proper is divided into six chapters. After a detailed discussion of probability, it displays most of the classical statistical tools, always with examples drawn from language study, using in most cases data taken from the literature on language and style or collected by the author and his co-workers. Following each section of the volume, with the exception of the later sections of chapter 6, are exercises which reinforce the material in the section or extend it. The aim has been to provide essential background for students of linguistics and the humanities, at the advanced undergraduate or graduate level, in a volume that could be covered in a one-year course meeting two hours each week, or, if chapter 6 is omitted, in a half-year course meeting three hours per week.

The first five chapters are more or less traditional except for the inclusion of three of the more popular non-parametric tests in chapter 5, Hypothesis Testing. For each of the tests, an attempt has been made to indicate its limitations. Examples from the literature are given which illustrate where careless statements made about these limitations in some popular statistical texts have led to the misuse of some of the tests. The sixth chapter deals in an extended manner with certain specific problems in the statistics of style and in lexicostatistics.

The design of the volume has not made it possible to include a specific discussion of every statistical index and technique used by linguists and stylisticians. There are good handbooks of statistical tests already available (see Bibliography at the end of the book), which serve this purpose well. Nor has it been possible to consider all aspects in which the reader might be interested. Some of the more interesting statistical studies are, however, mentioned in section 6.5, and various applications of statistics to historical linguistics are considered (see especially section 6.4). For those wishing to obtain information about stylistic studies in general, I would draw attention to the bibliographies of Bailey and Burton (1968) and Milic (1967a), as well as the collections of essays edited by Chatman (1971) and Sebeok (1960). In the area of linguistics, using the narrow sense of the term as the study of natural language (as opposed to literary text), the application of statistics (and mathematics in general, for that matter) has received little attention. Some exceptions are the application to phonetics and psycholinguistics as can be seen in the pages of the journals *Language and Speech* and *Journal of Verbal Learning and Verbal Behavior* and to a lesser extent applications to linguistic geography and dialectology as found in, for example, Greenberg (1956) and Houck (1967). A reading of Labov (1966) with statistics in mind should indicate some possibilities in this area. Those interested in applications of mathematics (and not statistics) to linguistics should see Brainerd (1971).

I wish to thank my colleagues D.B. DeLury and Ralph Wormleighton for their encouragement and stimulating conversation on the material and my co-workers Evelyn Center, Victoria Neufeldt, and Rubin Friedman for their help in collecting data, proofreading, and performing calculations. I also wish to thank the National Research Council of Canada, whose grant A5252 supported some of the essential costs incurred in production of the book.

This work has been published with the aid of the Publications Fund of the University of Toronto Press.

Toronto                                                 B.B.
December 1972

WEIGHING EVIDENCE IN LANGUAGE AND LITERATURE

A statistical approach

# Mathematical preliminaries

The reader is assumed to have already obtained at least a limited mathematical background before he attempts to read this book. This background should consist of a knowledge of elementary theory of sets, some knowledge of the properties of numbers, and an intuitive grasp of the concept of limit. The major definitions and results from these areas, which will be used without mention, are outlined below.

We have attempted to provide a treatment that can be understood without a prior knowledge of the calculus. However, some technical arguments involving the calculus are included, mainly for those who have some knowledge of this subject. We have tried to include sufficient explanation to ensure that the results of the technical arguments can be easily understood by readers who have no experience in the calculus.

## 0.1
### ELEMENTARY SET THEORY

The definitions and results of elementary set theory which we shall need are not deep and require no profound philosophical ideas. For us a *set* is a collection of objects, the *elements* of the set. We adopt the convention that *any set is uniquely determined by the elements it contains*. This means that a set $A$ is defined or determined if we can answer the question 'Does $x$ belong to $A$?' for any object $x$.

This relation of 'belongs to' is fundamental to the study of sets and is given the abbreviated form '$\in$.' Thus '$x$ belongs to $A$' is rendered '$x \in A$.' We also use '$x \notin A$' to stand for '$x$ does not belong to $A$.'

Sets can be described in a number of different ways: (1) by listing the elements of the set between braces, e.g., $\{1, 2, 3, 4, 5\}$; (2) by describing them in words, e.g., 'the first five natural numbers'; (3) by writing a symbolic description using braces, e.g., $\{x \mid x \in N \text{ and } x \leq 5\}$, which is read 'the set of $x$ such that $x$ belongs to $N$ (the set of natural numbers) and $x$ is less than or equal to 5.'

Since the *natural numbers* constitute the set $N$ of numbers used for counting (i.e., the set $N$ contains the numbers

1, 2, 3, 4, 5, 6, ...

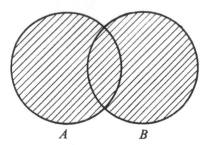

Shaded area represents
$A \cup B$

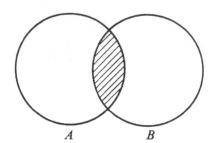

Shaded area represents
$A \cap B$

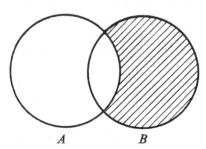

Shaded area represents
$B - A$

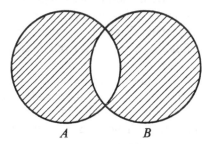

Shaded area represents
$A \oplus B = (B - A) \cup (A - B)$

FIGURE 0.1

where the three dots (...) are a short-hand way of saying etc., etc.), it should be clear that the three examples listed above are three descriptions of the same set.

Among the definitions and results about sets that we shall use without comment in the sequel are the following:

(a) $\{x \mid P(x)\}$ is the set of objects $x$ which satisfy some defining property, which we abbreviate $P(x)$.

(b) $\{x \in S \mid P(x)\}$ is the set of $x$ belonging to the set $S$ which satisfy $P(x)$.

(c) Two sets $A$ and $B$ are equal, $A = B$, provided $x \in A$ if and only if $x \in B$.

(d) $A$ is a *subset* of $B$ $(A \subseteq B)$ if $x \in A$ implies that $x \in B$.

(e) Thus $A = B$ if and only if $A \subseteq B$ and $B \subseteq A$.

(f) $A$ is a *proper* subset of $B$ $(A \subsetneqq B)$ if $A \subseteq B$ and $A \neq B$.

(g) The *union* of two sets $A$ and $B$ is the set $A \cup B = \{x \mid x \in A$ or $x \in B\}$ where 'or' means $x \in A$ or $x \in B$ or both (see figure 0.1).

(h) The *intersection* of two sets $A$ and $B$ is the set $A \cap B = \{x \mid x \in A$ and $x \in B\}$ (see figure 0.1).

(i) The *relative complement of $A$ with respect to $B$* is the set $B - A = \{x \mid x \in B$ and $x \notin A\}$ (see figure 0.1).

(j) The *symmetric difference* of $A$ and $B$ is $A \oplus B = (B-A) \cup (A-B)$ (see figure 0.1).

(k) Two special sets are important to consider: the *universe of discourse* in a given situation, say $U$, which contains as subsets all sets of a certain type under discussion, and the *empty set* $\emptyset$ which contains no elements and can be thought of as a subset of every set. It is easy to show that $U - \emptyset = U$ and $U - U = \emptyset$.

(l) Sometimes we shall need to discuss $2^U$, the *set of all subsets of $U$*; this includes $U$ itself and the empty set. Thus if $U = \{0, 1\}$, then $2^U$ contains exactly the sets $U$, $\{0\}$, $\{1\}$, $\emptyset$. In general, if $U$ is a finite set containing $n$ elements, then $2^U$ contains $2^n$ elements; hence the notation $2^U$.

(m) If $A$ and $B$ are sets, then their *cartesian product* is $A \times B = \{(a, b) \mid a \in A$ and $b \in B\}$ where $(a, b)$ is the *ordered pair* with first entry $a$ and second entry $b$. If $A = B$ and $a \neq b$, then of course $(a, b) \neq (b, a)$. We shall have cause to speak of cartesian products of more than two factors. Thus if $A_1, A_2, \ldots, A_n$ are sets,

$$A_1 \times A_2 \times \ldots \times A_n = \{(a_1, a_2, \ldots, a_n) \mid a_i \in A_i, i = 1, 2, \ldots, n\}$$

is their *cartesian product*. $(a_1, a_2, \ldots, a_n)$ is called an ordered *n*-tuple.

(n) The subsets of a set $U$ form a *Boolean algebra* with respect to the set-operations $\cup$, $\cap$, and complement $'$ where $A' = U - A$, which is another way of saying that the following conditions are satisfied for subsets $A$, $B$, and $C$ in $U$:

B1a $\quad A \cup B = B \cup A$
B1b $\quad A \cap B = B \cap A$
B2a $\quad (A \cup B) \cup C = A \cup (B \cup C)$
B2b $\quad (A \cap B) \cap C = A \cap (B \cap C)$
B3a $\quad A \cap (B \cup C) = (A \cap B) \cup (A \cap C)$
B3b $\quad A \cup (B \cap C) = (A \cup B) \cap (A \cup C)$
B4a $\quad A \cup \emptyset = A$
B4b $\quad A \cap U = A$
B5a $\quad A \cup A' = U$
B5b $\quad A \cap A' = \emptyset$

(o) A *function f from a set A to a set B* is a rule which associates with each $x \in A$ a unique element $f(x) \in B$. $A$ is the *domain* of $f$ and $B$ its *counter-domain*. We sometimes say that $f$ is *defined on $A$* when $A$ is its domain.

(p) A function $f$ from $A$ to $B$ is *one-to-one onto $B$* if for each $y \in B$ there is exactly one $x$ in $A$ such that $y = f(x)$.

(q) A function $f$ from $A$ to $B$ is completely determined by the set $\{(x, f(x)) \in A \times B \mid x \in A\}$, which is called the *graph of $f$* and can be identified with the function.

The most important example of a cartesian product is $R \times R$ where $R$ represents the set of real numbers. (They are considered in more detail in section 0.2.) $R \times R$ is sometimes called $R^2$, using the analogue of the cartesian product with

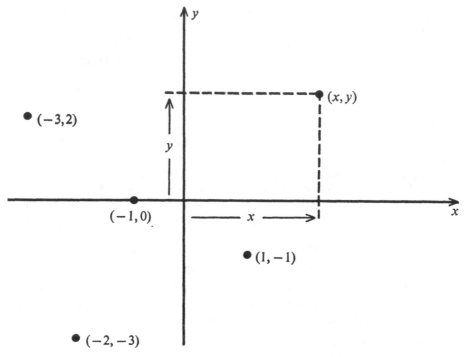

FIGURE 0.2

ordinary multiplication, and is in fact the *Euclidean plane* of ordinary analytic geometry. In analytic geometry a reference point, the *origin*, is chosen together with x- and y-axes as in figure 0.2. Each ordered pair (x, y) of real numbers represents exactly one point on this plane, i.e., the point exactly x units over to the right of and y units up from the origin as in figure 0.2. Of course, negative x- and y-values mean that the distances are measured in the opposite directions. Note the locations of $(-1, 0)$, $(1, -1)$, $(-3, 2)$, $(-2, -3)$ in figure 0.2.

We shall find that one of the most important kinds of function that we must consider is the *linear function* from $R$ to $R$ defined by an equation of the form

(1)  $f(x) = ax + b$

for fixed real values $a$, $b$. The graph of this function is a straight line in $R^2$ passing through the points $(0, b)$ and $(-b/a, 0)$, as indicated in figure 0.3. A little calculation will show that if $(x_1, y_1)$ and $(x_2, y_2)$ are two distinct points on the graph of the function defined by (1), then

(2)  $\dfrac{y_2 - y_1}{x_2 - x_1} = a$

for all choices of such points; the left side of (2) defines the *slope* of the line.

Next in importance is the *quadratic function* defined by an equation of the form

(3) $f(x) = ax^2 + bx + c$

where $a$, $b$, and $c$ are again fixed real numbers. The graph of such a function is usually a parabola, whose form depends on the choice of $a$, $b$, and $c$. I say *usually* a parabola because choosing $a = 0$ reduces (3) to a linear expression whose graph, as we have already said, is a straight line.

In figure 0.4 we show the graph of the function given by (3) when $a = 1$ and $b = c = 0$.

For more on these concepts see volume I, chapter 1 and volume III, chapter 1 of Brainerd et al. (1967), Suppes (1957), or Kemeny, Snell, and Thompson (1957).

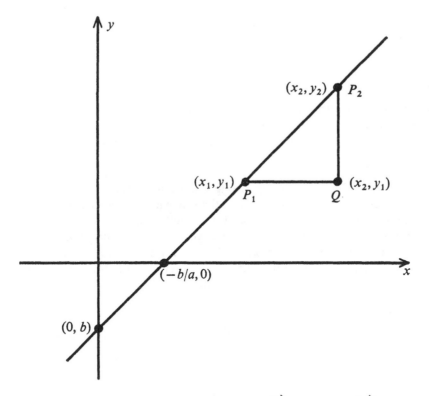

FIGURE 0.3   Graph of $\{(x, ax+b) \mid x \text{ in } R\}$. The rise, $\overrightarrow{QP_2}$, over the run, $\overrightarrow{P_1Q}$, i.e.,

$\dfrac{\overrightarrow{QP_2}}{\overrightarrow{P_1Q}} = \dfrac{y_2 - y_1}{x_2 - x_1}$ , is the slope.

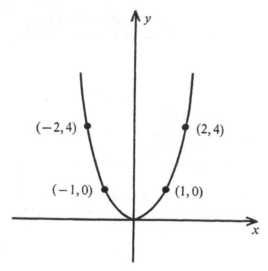

FIGURE 0.4   Graph of the parabola $\{(x, x^2) \mid x \text{ real}\}$.

## 0.2
### NUMBER SYSTEMS

We shall need to deal with four number systems explicitly in the course of our discussions of statistics. The first and simplest is the set of *natural numbers*, usually called $N$ and composed of the numbers used for counting:

$$1, 2, 3, 4, \ldots$$

We normally think of this set as supporting the usual operations of addition and multiplication and the relation $\leqq$, less than or equal to, so that $n \leqq m$ if either $n = m$ or there is a natural number $k$ such that $n+k = m$.

Not all equations of the form $x+a = b$ have solutions $x$ when $a$ and $b$ are arbitrary natural numbers. For example $x+3 = 2$ has no solution $x \in N$. Therefore, to provide such equations with solutions, we extend the natural numbers to the integers. The set $I$ of *integers* is composed of

$$0, \pm 1, \pm 2, \pm 3, \ldots$$

where the natural numbers are identified with the integers prefaced by a plus sign in a natural way:

$$1 \leftrightarrow +1,$$
$$2 \leftrightarrow +2,$$
$$\cdots$$
$$n \leftrightarrow +n,$$
$$\cdots$$

so that we can write $N \subsetneqq I$, i.e. $N$ is a proper subset of $I$.

In the integers we can add, multiply, and subtract so that $x+a = b$ has a solution $x = b-a$ for any pair of arbitrary integers $a$, $b$. We cannot always divide, however; i.e., the equation $ax = b$ cannot always be solved in the integers. Nevertheless, the integers can be extended to the *rational numbers*

$$Q = \{p/q \mid p \in I, q \in N\},$$

and in $Q$ the equation $ax = b$ can be solved uniquely for all cases except $a = 0$. In this case, $0 \cdot x = 0$ for all $x \in Q$. The integers are identified in $Q$ with those fractions of the form $p/1$ as follows:

$$0 \leftrightarrow 0/1,$$
$$+1 \leftrightarrow +1/1,$$
$$-1 \leftrightarrow -1/1,$$
$$\cdots$$

(Note that $p/q = p_1/q_1$ if and only if $pq_1 = p_1q$; for example consider $1/2 = 4/8$.) Then $I$ is a proper subset of $Q$.

The rational numbers have certain shortcomings which make it necessary to extend them as well. In particular, not every subset $A$ of $Q$ which has an *upper bound $m \in Q$* ($m$ is an *upper bound* of $A$ if $a \leq m$ for all $a \in A$) has a least upper bound in $Q$. This lack of a rational least upper bound for the set $\{x \in Q \mid x^2 < 2\}$ allows the equation $x^2 = 2$ to have no rational solutions. The extension of $Q$ obtained by adding these least upper bounds is called $R$, the set of *real numbers*, which we usually depict as the number line (see figure 0.5).

The advantage of the real numbers over the rationals is that *every set $A$ in $R$ which has an upper bound has a least upper bound.* This least upper bound of $A$ is usually denoted sup $A$ and is by definition the smallest real number greater than or equal to every element of $A$.

As an example of a subset of $Q$ with a least upper bound consider the set $\{(n-1)/n \mid n = 1, 2, 3, \ldots\}$. Any number greater than or equal to 1 is an upper bound and 1, the smallest of these, is the least upper bound.

As it turns out, every real number has a decimal representation, and those real numbers with ultimately repeating decimal expressions like $0.333\ldots$ or $0.123123123\ldots$ are exactly the rational numbers; thus $Q$ is a proper subset of $R$.

We have, then, the following hierarchy of number systems:

$$N \subsetneq I \subsetneq Q \subsetneq R.$$

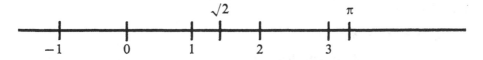

FIGURE 0.5

There is one other system, the complex numbers, which, though it is an important number system, we have no need of here.

For more on these number systems consult Courant and Robbins (1941).


## 0.3
### SEQUENCES

An important class of functions is that composed of the functions from $N$ or $N_0 = N \cup \{0\}$ to $R$. These are called *sequences*. Thus, if $f$ is a sequence, $f(n)$ is a real number for each $n \in N$ (or $n \in N_0$ as the case may be). We usually denote a sequence $f$ where $f(n) = x_n$ by the symbol

$$\{x_n\}_{n=0}^{\infty}$$

when the domain of $f$ is $N_0$, and by the symbol

$$\{x_n\}_{n=1}^{\infty}$$

when the domain of $f$ is $N$, the superscript and subscript indicating the limits of the domain of the function. In contexts where the domain of $f$ is understood to be fixed as one or the other, we drop superscript and subscript to the brackets and write simply $\{x_n\}$.

At times it is useful to simply list the elements of the sequence. Thus if $x_n = 1 - 1/n$, then

$$\{x_n\}_{n=1}^{\infty} = \left\{ 0, \frac{1}{2}, \frac{2}{3}, \frac{3}{4}, \ldots, 1 - \frac{1}{n}, \ldots \right\}.$$

Our primary concern when discussing sequences is their behaviour as $n$ becomes large (as $n \to \infty$, or as $n$ *approaches infinity*). Three kinds of behaviour will be of interest in particular: (1) $x_n$ approaches a fixed number $L$ as $n \to \infty$; (2) $x_n$ becomes large without limit as $n \to \infty$ (this is written $x_n \to \infty$); (3) $x_n$ becomes negative without limit as $n \to \infty$ (this is written $x_n \to -\infty$). Other kinds of behaviour such as oscillation are also observed: for example $\{(-1)^n\}$, $\{(-1)^n n\}$, and $\{n + (-1)^n n\}$ are sequences that oscillate as $n \to \infty$.

The following three sequences illustrate (1), (2), and (3) respectively:

$$\{(n+1)/n\}, \quad \{n^2\}, \quad \{-n^3\}.$$

The sequence $\{(n+1)/n\}$ clearly approaches 1 in the sense that whatever tolerance $\varepsilon > 0$ one chooses, there is an $n_\varepsilon \in N$ (depending on the $\varepsilon$ chosen) such that for all $n > n_\varepsilon$

$$1 - \varepsilon < (n+1)/n < 1 + \varepsilon.$$

Thus if one 'goes out' far enough in the sequence $\{x_n\} = \{(n+1)/n\}$ the difference, $|1 - x_n|$, between 1 and $x_n$ will from that point onwards be smaller than $\varepsilon$.

This situation is a special instance of the following definition of a *limit*:

$$\lim_n x_n = L$$

if and only if for each positive real number $\varepsilon > 0$ there is a natural number $n_\varepsilon$ such that if $n > n_\varepsilon$, then

$$|L - x_n| < \varepsilon.$$

The expression $\lim_n x_n = L$ is read 'the limit of the sequence $\{x_n\}$ as $n \to \infty$ is $L$.'

The sequence $\{n^2\}$ becomes large without limit as $n \to \infty$ in the sense that for any positive real number $M$, there is an $n_M \in N$ (depending on $M$) such that if $n > n_M$, then the corresponding term $x_n = n^2$ will be larger than $M$, i.e., $n^2 > M$ whenever $n > n_M$. For example, if $M = 10{,}000$, then $n_M$ could be 100. Indeed if $n > 100$, then $n^2 > (100)^2 = 10{,}000$ so if $n > 100$, $x_n = n^2 > M$.

In general, we say that $x_n \to \infty$ as $n \to \infty$ or $\lim_n x_n = \infty$ if and only if for any real $M > 0$ there is an $n_M \in N$ such that if $n > n_M$, then $x_n > M$.

Similarly we say that $x_n \to -\infty$ as $n \to \infty$ or $\lim_n x_n = -\infty$ if and only if for any real $M > 0$ there is an $n_M \in N$ such that if $n > n_M$, then $x_n < -M$.

Among the properties of sequences, the following are fundamental. If $\lim_n x_n = L$, $\lim_n y_n = M$, and $L$, $M$, and $a$ are real numbers (in particular not $\pm \infty$), then

(1) $\lim_n ax_n = aL$,

(2) $\lim_n (x_n + y_n) = L + M$,

(3) $\lim_n (x_n y_n) = LM$,

and if $M \neq 0$,

(4) $\lim_n (x_n / y_n) = L/M$.

For more information on sequences, see chapter 1 of volume II of Brainerd et al. (1967) or J.A. Green (1958).

## 0.4
### CONTINUITY AND DIFFERENTIATION

Another kind of function which is of fundamental importance is the real-valued function defined on some subset of the real number line. Usually this subset will be one of the following three kinds of set:

(i) A finite interval which is *open*, i.e., of the form

$$(a, b) = \{x \in R \mid a < x < b\};$$

*semi-open*, i.e., either of the form

$$[a, b) = \{x \in R \mid a \leq x < b\}$$

or of the form

$$(a, b] = \{x \in R \mid a < x \leqq b\};$$

or *closed*, i.e., of the form

$$[a, b] = \{x \in R \mid a \leqq x \leqq b\}.$$

(ii) The *non-negative half-line*

$$[0, \infty) = \{x \in R \mid 0 \leqq x\}.$$

(iii) The entire *real line R*.

Suppose $f$ is a real-valued function defined on some open interval $(a, b)$ and $c$ is a point in the (closed) interval $[a, b]$. We say that

$$\lim_{x \to c} f(x) = L$$

if for each $\varepsilon > 0$ there is a $\delta > 0$ (depending on $\varepsilon$ of course) such that if $x \in (a, b)$ and $|x - c| < \delta$, then

$$|f(x) - L| < \varepsilon.$$

Thus if $f$ has domain $(0, 1)$ and is given by the rule $f(x) = x$, then

$$\lim_{x \to 0} f(x) = 0, \quad \lim_{x \to \frac{1}{2}} x = \tfrac{1}{2}, \quad \lim_{x \to 1} x = 1.$$

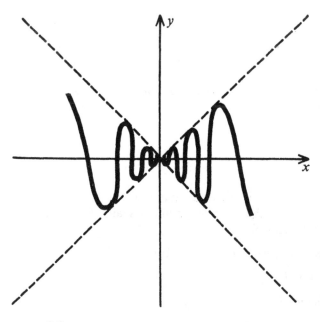

FIGURE 0.6

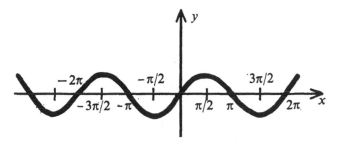

FIGURE 0.7   Graph of the function sin = {(x, y) | y = sin x}.

A more interesting example is the function, call it g, defined on (−1, 1) by the rule

$$g(x) = x \sin(1/x) \quad \text{for } x \neq 0,$$

$$= 0 \qquad\quad \text{for } x = 0.$$

This function oscillates toward 0 as $x \to 0$, as can be seen from figure 0.6. Thus

$$\lim_{x \to 0} [x \sin(1/x)] = 0$$

even though the expression $x \sin(1/x)$ can give no value for $g(0)$. Remember that 1/0 is undefined. The graph of the sine function $y = \sin x$ is given in figure 0.7. Its values for specific values of $x$ can be obtained from trigonometric tables.

Limits here have similar properties to limits of sequences. If $f$ and $g$ are defined at least on an interval $(a, b)$ and $c \in [a, b]$, then if the limits on the right exist,

(1)  $\lim_{x \to c} [f(x) + g(x)] = \lim_{x \to c} f(x) + \lim_{x \to c} g(x),$

(2)  $\lim_{x \to c} [f(x) \cdot g(x)] = \lim_{x \to c} f(x) \lim_{x \to c} g(x);$

if $e$ is a real number,

(3)  $\lim_{x \to c} e \cdot f(x) = e \lim_{x \to c} f(x);$

and if $g(x) \neq 0$ in $(a, b)$ and $\lim_{x \to c} g(x) \neq 0$, then

(4)  $\lim_{x \to c} \dfrac{f(x)}{g(x)} = \dfrac{\lim_{x \to c} f(x)}{\lim_{x \to c} g(x)}.$

A function $f$ whose domain includes $(a, b)$ is *continuous at a point* $c \in (a, b)$ provided

$$\lim_{x \to c} f(x) = f(c).$$

If $f$ is continuous at every point in $(a, b)$ in the above sense, we say that $f$ is *continuous on the interval* $(a, b)$. Continuity of $f$ on $(a, b)$ corresponds to the intuitive notion that the graph of $f$ has no break at any $x$ in $(a, b)$; see figure

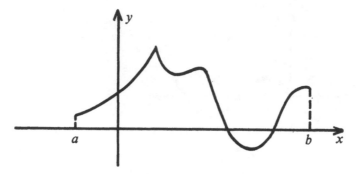

FIGURE 0.8   The function with graph $\{(x, y) \in R \times R \mid y = f(x)\}$ is continuous in $(a, b)$.

0.8. If $f$ is not continuous at a point $c$, we say it is *discontinuous* at $c$ or has a *discontinuity at c*. See figures 0.9, 0.10, and 0.11 for examples of discontinuities at various points.

Most of the real-valued functions considered here will be continuous. They will also be smooth enough to possess an instantaneous rate of change at all points in their domain of definition, in the sense made specific in the following paragraph.

Consider the function depicted in figure 0.12. The change in $f(x)$ between $x_0$ and $x_0 + h$ is given by the expression

$$\Delta_h f(x_0) = f(x_0 + h) - f(x_0),$$

and the *average rate of change* between $x_0$ and $x_0 + h$ is, by definition,

$$\frac{\Delta_h f(x_0)}{h} = \frac{f(x_0 + h) - f(x_0)}{h}$$

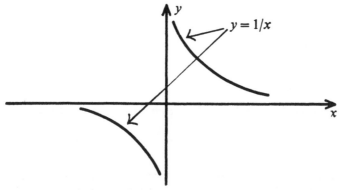

FIGURE 0.9   The function with graph $\{(x, y) \mid y = 1/x\}$ has an infinite discontinuity at $x = 0$.

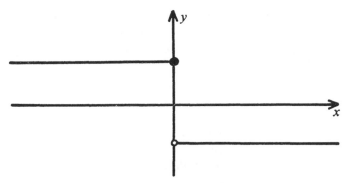

FIGURE 0.10   The function with graph $\{(x, y) \mid y = 1 \text{ for } x \geq 0,$
$y = -1 \text{ for } x < 0\}$ has a finite discontinuity at $x = 0$.

As $h$ approaches zero, we obtain a 'better' estimate of what might be termed the '(instantaneous) rate of change' of $f(x)$ at $x_0$. Thus we define the *rate of change of $f$ at $x_0$* to be

$$f'(x_0) = \lim_{h \to 0} \frac{\Delta_h f(x_0)}{h} = \lim_{h \to 0} \frac{f(x_0+h) - f(x_0)}{h},$$

which is often called the *derivative of $f$ at $x_0$* and is usually denoted $f'(x_0)$.

In most applications that we discuss, the important feature of this derivative $f'(x_0)$ will be that it is the slope of the tangent (straight) line to the graph of $f$

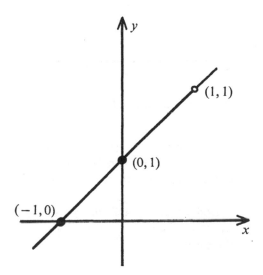

FIGURE 0.11   The function with graph $\left\{(x, y) \,\middle|\, y = \dfrac{x^2 - 1}{x - 1}\right\}$
is not defined when $x = 1$, and hence is discontinuous there.

at $x = x_0$ (see figure 0.12). The slope $m$ of a straight line is by definition (see section 0.1) the ratio of the change, $y_1 - y_0$, in the second coordinate to the corresponding change, $x_1 - x_0$, in the first coordinate between two points $(x_0, y_0)$ and $(x_1, y_1)$ on the line. Thus

$$m = \frac{y_1 - y_0}{x_1 - x_0}.$$

The main application of the derivative lies in the fact that if a function $f$ is defined on some (open) interval, $(a, b)$ say, and has a derivative at every point in that interval, and if it has a maximum or minimum value, this value will occur

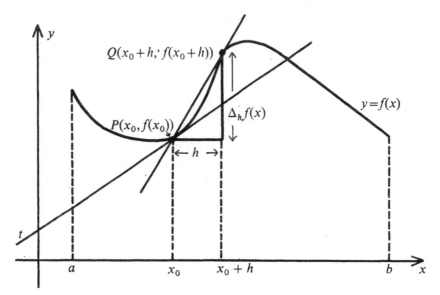

FIGURE 0.12   As $h \to 0$ the secant line $\overline{PQ}$ approaches the tangent line $t$ at $P$.

for an $x_0$ for which the tangent line to the graph of $f$ at $(x_0, f(x_0))$ is horizontal, and hence its slope is 0. Thus at such a point $f'(x_0) = 0$. Therefore to find any such maximum or minimum values, we need only consider those values of $x$ for which $f'(x) = 0$ (see figure 0.13).

On rare occasions we shall need to consider maxima and minima of real-valued functions defined on $R^n$, the cartesian product of $n$ copies of the real number line, i.e., functions with rules of the form

$$y = f(x_1, x_2, \ldots, x_n).$$

These maxima and minima occur only at points $(x_1{}^0, x_2{}^0, \ldots, x_n{}^0) \in R^n$ where all the *partial derivatives*

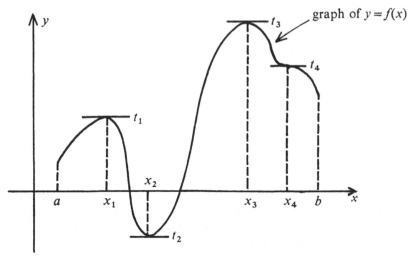

FIGURE 0.13   The slopes of the tangents $t_1$, $t_2$, $t_3$, $t_4$ to the graph of $y = f(x)$ at $(x_1, f(x_1))$, $(x_2, f(x_2))$, $(x_3, f(x_3))$, and $(x_4, f(x_4))$ respectively are all zero because $f'(x_1) = f'(x_2) = f'(x_3) = f'(x_4) = 0$. The maximum value of $f(x)$ (and the minimum too, for that matter) occurs at one of the points where $f'(x) = 0$, in this case.

$$\frac{\partial f(x_1{}^0, ..., x_n{}^0)}{\partial x_1} = f_{x_1}(x_1{}^0, ..., x_n{}^0)$$

$$= \lim_{h \to 0} \frac{f(x_1{}^0 + h, x_2{}^0, ..., x_n{}^0) - f(x_1{}^0, x_2{}^0, ..., x_n{}^0)}{h},$$

$$\frac{\partial f(x_1{}^0, ..., x_n{}^0)}{\partial x_2} = f_{x_2}(x_1{}^0, x_2{}^0, ..., x_n{}^0)$$

$$= \lim_{h \to 0} \frac{f(x_1{}^0, x_2{}^0 + h, ..., x_n{}^0) - f(x_1{}^0, x_2{}^0, ..., x_n{}^0)}{h},$$

...

$$\frac{\partial f(x_1{}^0, ..., x_n{}^0)}{\partial x_n} = f_{x_n}(x_1{}^0, x_2{}^0, ..., x_n{}^0)$$

$$= \lim_{h \to 0} \frac{f(x_1{}^0, x_2{}^0, ..., x_n{}^0 + h) - f(x_1{}^0, x_2{}^0, ..., x_n{}^0)}{h}$$

equal zero.

There are some very well developed techniques for finding derivatives and partial derivatives which can be found in any book on the calculus. See volume II of Brainerd et al. (1967) or Hilton (1958) for example. Readers interested can read further, but the remarks above should be sufficient to achieve an understanding of the development which follows.

# 1
# Reduction of numerical data

The word statistics can be construed in a number of different ways. We shall be concerned here particularly with two of these meanings: (1) the collecting, organizing, summarizing, and analysing of quantitative information; (2) a set of techniques for drawing inferences and generalizations from small collections, called *samples*, to larger collections, called *populations*, using the mathematical theory of probability.

Meaning (1) is the subject of this chapter, and (2) will be discussed in some detail in later chapters. The distinction between (1) and (2) can be made manifest by calling the statistics defined in (1) *descriptive statistics* and that defined in (2) *inferential statistics*.

We shall find that inferential statistics provides a collection of techniques for making decisions in the face of uncertainty. For example, when the author of a work of literature is known to be one of $A_1$, $A_2$, ..., $A_n$ but it is not known which, then the results of inferential statistics can sometimes be employed in order to show that, given the information at hand, it is significantly more likely that a particular one, say $A_1$, wrote the piece than any of the others. This is not to say that $A_1$ actually wrote the piece, because the information used may have been misleading. It says only that, given the information, $A_1$ is the most likely author and our chances of being wrong are smaller than a certain amount.

The fundamental aim of descriptive statistics is to present the data in an explicit and informative way, in order to facilitate their communication and interpretation. This process of bringing order out of raw data often aids in the construction of hypotheses that lead to mathematical models of the mechanism producing the data. These descriptive techniques are the main concern of this chapter.

## 1.1
MEASUREMENTS AND THE ARITHMETIC MEAN

Any scientific study of an aspect of human behaviour is based on observations of that behaviour. This will, of course, be true also of scientific studies in linguistics and literature. Just as in psychology where the observations are made upon individual subjects, in studies of language and style the observations are usually made upon corpora of utterances made by individuals. For example, we might observe in a given set of corpora the ratio of the number of nouns to the total number of words in each corpus, or we might consider the ratio of the number of word-*types* (different words) used in the corpus to the number of word-*tokens* (words) used. For example, the sentence

*The boy gave the other boy his ball*

contains eight word-tokens; however, there are only six word-types represented:

*the, boy, gave, other, his, ball.*

Thus the ratio in question would be

$$\frac{\text{no. of word-types}}{\text{no. of word-tokens}} = \frac{6}{8} = \frac{3}{4},$$

Another situation that might arise is that in which a researcher may wish to ascertain the relative frequency of the phoneme /p/ in some dialect of present-day English. In order to obtain his result, he might sample a large number of speakers of this dialect chosen at random. From each speaker $S_i$ he would obtain a corpus $C_i$ from which he could calculate the ratio

$$\frac{\text{no. of instances of /p/ in } C_i}{\text{no. of sound-tokens in } C_i} = \frac{n_i(\text{p})}{n_i} = f_i.$$

If he combined all his corpora to form one large corpus[1]

$$C_1 \cup C_2 \cup ... \cup C_m = C,$$

then he would obtain the relative frequency

$$f = \frac{\text{no. of instances of /p/ in } C_1 \cup C_2 \cup ... \cup C_m}{\text{no. of sound-tokens in } C_1 \cup C_2 \cup ... \cup C_m}$$

$$= \frac{n_1(\text{p}) + n_2(\text{p}) + ... + n_m(\text{p})}{n_1 + n_2 + ... + n_m}$$

$$= \frac{n_1 f_1 + n_2 f_2 + ... + n_m f_m}{n_1 + n_2 + ... + n_m}.$$

If all the individual corpora $C_i$ were of the same length, say $n$ phonemes long, then

$$f = \frac{n f_1 + ... + n f_m}{m \cdot n} = \frac{f_1 + f_2 + ... + f_m}{m}$$

---

1  See section 0.1 for the definition of $\cup$.

The expression

$$\frac{f_1 + \dots + f_m}{m},$$

composed of the sum of a number $m$ of observations $f_i$ of some variable divided by $m$, appears often in studies conducted by behavioural scientists and is called the *arithmetic mean* of the observations $f_1, \dots, f_m$.

In general, we are concerned with observing attributes of some fixed possibly infinite class of corpora, the *population*. The attributes are assumed to manifest themselves as numerical quantities and may vary from corpus to corpus within the population. Such numerical attributes are called *variates*. Sometimes the variation of a given variate is systematic, sometimes it is due to chance, and sometimes it is due to a combination of both. Part of our job is to determine how much of this variation is due to chance.

Usually, it is impossible to obtain information about the whole population; only information about small finite subsets of the population can be obtained. For example we may choose corpora $C_1, C_2, \dots, C_n$ in that order from a population of corpora. This ordered set $(C_1, C_2, \dots, C_n)$ is called a *sample of the population*. If for each $C_i$ we compute the value $x_i$ that a variate $X$ takes in $C_i$, then the ordered set $(x_1, x_2, \dots, x_n)$ is called a *sample of the variate X*. For example, if the population is the set of descriptive paragraphs written by a certain author, we might choose a sample of $n$ descriptive paragraphs $(P_1, P_2, \dots, P_n)$ at random from his works and compute in each case the value $x_i$ of the variate

$$X = \frac{\text{no. of noun-tokens in the paragraph}}{\text{no. of word-tokens in the paragraph}}.$$

Then $(x_1, \dots, x_n)$ is the sample of the variate $X$ corresponding to $(P_1, \dots, P_n)$.

In general the *arithmetic mean* of a sample $(x_1, x_2, \dots, x_n)$ of a variate $X$ is defined to be

(1)  $\bar{x} = \dfrac{x_1 + x_2 + \dots + x_n}{n}.$

The arithmetic mean of $x_1, x_2, \dots, x_n$ is a measure of the centre of gravity of the distribution of the values $x_i$. Indeed, if for each $x_i$ a particle $p_i$ of mass one is placed on a line at a distance $x_i$ from a fixed origin $O$ (as in figure 1.1) and the resulting configuration of particles is held together by a rigid though weightless frame, then a knife edge placed under the frame at a distance

$$\bar{x} = (x_1 + x_2 + \dots + x_n)/n$$

to the right of $O$ in this case will support the frame in perfect equilibrium. The point-masses to the left of $\bar{x}$ supply a counter-clockwise torque about $\bar{x}$ which

FIGURE 1.1   The configuration of the particle-masses when, in the sample $(x_1, x_2, \ldots, x_n)$, the sample values are listed in increasing order with $i$, i.e., $x_1 < x_2 < \ldots < x_n$.

exactly equals the clockwise torque supplied by the point-masses on the right of $\bar{x}$. The point $\bar{x}$ is called the centre of mass or centre of gravity of the configuration particle $p_1, \ldots, p_n$.

For a researcher wishing to find the relative frequency of /p/ in a certain dialect, the rationale is that the more people he samples the closer this centre of gravity,

$$\bar{f} = \frac{f_1 + \ldots + f_m}{m},$$

will come to the actual relative frequency.[2]

EXAMPLE 1   In five samples each composed of the first 50 characters of chapters 1, 3, 10, 27, and 31 of the *Tao Teh Ching* (the text used is the edition appearing as Asian Institute Translations No. 1) we find that 29, 36, 36, 34, and 31 different Chinese characters appear, respectively. Thus the average number of different characters appearing in these 50-character stretches of corpus is

$$\bar{x} = \tfrac{1}{5}(29 + 36 + 36 + 34 + 31) = 33.2.$$

On two occasions we have informally used the term 'random' in relation to a choice of sample. We shall now consider this term in more detail. In order to obtain a truly representative sample $(C_1, C_2, \ldots, C_n)$ of a population, we must make sure that (i) each time a $C_i$ is chosen, every corpus in the population has the same chance of being chosen to be that $C_i$ as any other corpus, and (ii) the successive choices have no effect on one another, i.e., they are independent of one another. A sample satisfying (i) and (ii) may be called a *simple sample*.

In most books on statistics a *random sample* from a population is defined to be an $n$-tuple of individuals $(i_1, i_2, \ldots, i_n)$ chosen so that every such $n$-tuple is as likely to be chosen as any other. It should be observed that we have not ruled out the choice of the sample $(i_1, i_2, \ldots, i_n)$ where all the entries are equal. Later, in section 2.4, we shall consider another kind of random sampling where all the entries must be different. To highlight this distinction, the present definition of random sampling is sometimes called *random sampling with replacement* whereas

2   In some cases (this one in particular) this actual value is in the nature of a theoretical construct and can never be fully achieved in nature.

*B*

that to be discussed in section 2.4 is *random sampling without replacement*.

We shall see later (exercise 4 of section 2.4) that a random sample (with replacement) is a simple sample and vice versa, so we shall refer to the dual concept as a random sample.

This definition is meant to ensure that a random sample is, in so far as it is possible, representative or typical of the population from which it is drawn.

We illustrate a standard method for achieving randomness with the following example.

EXAMPLE 2  Suppose we wish to obtain the number of entries in a certain dictionary of, say, 750 full pages of entries, say pages 15–764 inclusive. If we did not need the exact number of entries but could settle for a reasonable estimate, we could proceed as follows. Select a sample of, say, six pages at random, count the number of entries on each page, obtain the average number of entries per page, and multiply this value by the number of pages containing entries, in this case 750. Our problem is then to obtain a random sample of six pages from pages 15–764 in the dictionary.

The most rigorous method might be to make 750 slips of paper, number them consecutively 15, 16, ..., 764, place them in a hat or urn, mix them thoroughly, draw one out, note down the number, replace the slip, mix again, and repeat the process five more times.

Another method is to resort to a *table of random numbers* in which the digits 0, 1, 2, ..., 9 are arranged in random order. A number of such tables have been published (see, for example, Rand 1955). For a job like this, however, one can generate a mini-table of random numbers by taking ten 3 by 5 inch cards, numbering them 0, 1, 2, ..., 9, shuffling them conscientiously, spreading them out numbered-side down, selecting one at random, and noting the number. Repeating the process, say, 30 times, we might obtain the sequence:

1, 5, 3, 2, 2, 7, 7, 5, 1, 5, 3, 7, 4, 2, 4, 8, 0, 3, 4, 7, 3, 5, 8, 2, 0, 8, 1, 7, 0, 2.

If we group these numbers in threes, we obtain

153, 227, 751, 537, 424, 803, 473, 582, 081, 702.

The first five fall into the required range (15–764), but the sixth does not. We discard it and pass to the next, 473, which falls in the range. Thus our sample pages are

153, 227, 751, 537, 424, 473.

On these pages we might obtain the following set of entry numbers: 40, 16, 32, 35, 34, 35. The mean number of words per page for this sample is

$$\bar{x} = \tfrac{1}{6}(40+16+32+35+34+35) = 32.$$

Therefore our estimate for the number of entries in the dictionary is

$$32 \times 750 = 24,000 \text{ words.}$$

See exercise 4 for another illustration of the use of a random-number table. For more on the use of random-number tables in literary studies see Yule (1944), chapter 3.

## EXERCISES

1 A number of the exercises in this chapter will be closely interconnected because of their dependence on the same data, which the reader may choose himself. This question is the first of these. The reader should choose for himself a corpus containing at least 100,000 words – a novel, for example, would be ideal. He can then use this corpus for the subsequent related questions.

(a) Choose at random in your corpus 10 continuous passages of exactly 50 words each. Count the number of different words in the passage and find the arithmetic mean of the 10 type-counts.

(b) Perform this experiment again using randomly selected passages of 25 words each.

(c) Perform (b) again but this time select the 25-word passages to be either the first or the last 25 words in each of the 50-word passages in part (a). Use a coin toss to make the decision.

2 Suppose $x_1, \ldots, x_n$ constitute a set of numbers and

$$\bar{x} = \frac{x_1 + x_2 + \ldots + x_n}{n}$$

is their mean.

(a) Show that if $y_i = kx_i$ where $k$ is a number, then

$$\bar{y} = (y_1 + \ldots + y_n)/n = k\bar{x}.$$

(b) If $y_i = x_i + a$ for some fixed value $a$, then show that $\bar{y} = \bar{x} + a$.

(c) If $z_1, z_2, \ldots, z_n$ is another set of numbers and $y_i = x_i + z_i$, then if $\bar{z} = (z_1 + z_2 + \ldots + z_n)/n$, show that $\bar{y} = \bar{x} + \bar{z}$.

3 Consider your results in exercise 1. Let $\bar{x}$, $\bar{y}$, and $\bar{z}$ stand for the means obtained from the results of parts (a), (b), and (c) respectively.

(a) How would you interpret $\bar{x}/50$, $\bar{y}/25$, and $\bar{z}/25$?

(b) Can you explain the difference among the values?

(c) If $\bar{w}_{100}, \bar{w}_{1000}, \bar{w}_{10,000}, \ldots$ were type-counts averaged over 100-word, 1000-word, 10,000-word, etc. stretches from a very long corpus in a given language, how would you imagine the sequence of values

$$\bar{w}_{100}/100, \bar{w}_{1000}/1000, \bar{w}_{10,000}/10,000, \ldots, \bar{w}_{10^k}/10^k$$

would behave as the number of words, $10^k$, in the stretch increased? Have you an explanation for this behaviour?

4 (a) Choose a dictionary and estimate the number of word-entries in it using a sample containing six pages selected at random by means of the following sequence of random digits taken from the Rand (1955) table, line 135, third column:

4, 3, 2, 5, 3, 8, 4, 1, 4, 5, 6, 0, 8, 3, 3, 2, 5, 9, 8, 3, 0, 1, 2, 9, 1, 4, 1, 3, 4, 9, 2, 0, 3, 6, 8, 0, 7, 1, 2, 6, 1, 4, 3, 8, 7, 0, 6, 3, 4, 5.

This sequence should be long enough to handle any dictionary with fewer than 1000 pages.

(b) If your dictionary contains more than 999 pages but less than, say, 2000 pages, how might you proceed using the sequence of digits presented so as to obtain a random sample of pages? What assumptions have you made about the random sequence?

5 Another measure of the centre of a sample $(x_1, x_2, ..., x_n)$ is its *median*, i.e., the middle one of the sample values $x_i$. For example, if the sample obtained in example 1 is arranged in ascending order of magnitude

29, 31, 34, 36, 36,

the middle one, in this case 34, is the median of the sample. If there is no middle value of the sample, then there is a $k$ such that when the sample is arranged in ascending order

$$x_1 \leqq x_2 \leqq ... \leqq x_n,$$

$x_1, x_2, ..., x_k$ constitutes half the sample values and $x_{k+1}, ..., x_n$ constitutes the other half. In this case the median is usually chosen to be $\frac{1}{2}(x_k + x_{k+1})$, a point mid-way between $x_k$ and $x_{k+1}$.

(a) If the sample contains an odd number of values $x_1 \leqq x_2 \leqq ... \leqq x_n$, write an expression for $k$ in terms of $n$ such that $x_k$ is the median.

(b) If the sample contains an even number of sample values, write $k$ in terms of $n$ so that $\frac{1}{2}(x_k + x_{k+1})$ is the median of this sample.

## 1.2
### SUMMATION NOTATION

In section 1.1 the arithmetic mean of the set $\{x_1, x_2, ..., x_n\}$ of numbers is given by the equation

$$\bar{x} = (x_1 + x_2 + ... + x_n)/n.$$

We often write this

$$\bar{x} = \frac{1}{n}\left(\sum_{k=1}^{n} x_k\right),$$

which is read '$\bar{x}$ equals one over $n$ times the summation of $x_k$ from $k$ equals one to $k$ equals $n$.'

In general, if

$$\{x_m, x_{m+1}, \ldots, x_{n-1}, x_n\}$$

is an indexed set of numbers (or any mathematical symbols for that matter), then by definition

$$x_m + x_{m+1} + \cdots + x_{n-1} + x_n = \sum_{k=m}^{n} x_k.$$

In particular, the special cases

$$x_k = \sum_{j=k}^{k} x_j, \qquad x_1 + x_2 = \sum_{l=1}^{2} x_l$$

should be observed.

In future we shall have cause to resort to this notational convention quite often, so it is well to know something of its capabilities. First note that

$$(1) \quad \sum_{k=m}^{n} x_k = x_m + x_{m+1} + \cdots + x_n = \sum_{j=m}^{n} x_j,$$

that is, the *index of summation*, in the first case $k$ and in the latter $j$, can be arbitrarily chosen. Since the resulting sum is independent of this choice, the index of summation is often called a *dummy* variable.

Other properties of this convention are listed below:

$$(2) \quad \sum_{i=1}^{n} (x_i + y_i) = \left( \sum_{i=1}^{n} x_i \right) + \left( \sum_{i=1}^{n} y_i \right).$$

Usually we assume that the sign of summation, $\sum$, takes precedence over ordinary '+' and write

$$\sum_{i=1}^{n} x_i + \sum_{i=1}^{n} y_i$$

for the right side of (2). This convention is analogous to that used unconsciously in standard arithmetic where $ax + b$ is written for $(ax) + b$, the other possible interpretation being $a(x + b)$. The precedence of multiplication over addition is always assumed when writing arithmetic expressions, i.e.,

$$(3) \quad \sum_{i=1}^{n} cx_i = c \sum_{i=1}^{n} x_i,$$

$$(4) \quad \sum_{i=1}^{n} a = na,$$

$$(5) \quad \sum_{i=1}^{n} \left( \sum_{j=1}^{m} a_{ij} \right) = \sum_{j=1}^{m} \left( \sum_{i=1}^{n} a_{ij} \right).$$

Like the parentheses in (2), the parentheses in (5) can be omitted.

(6) $$\left(\sum_{i=1}^{n} a_i\right)\left(\sum_{j=1}^{m} b_j\right) = \sum_{i=1}^{n}\sum_{j=1}^{m} a_i b_j = \sum_{j=1}^{m}\sum_{i=1}^{n} a_i b_j.$$

The analogues of expressions (2)–(6) when the limits of summation are different from 1 and $n$ can also be proved.

We shall prove (2), (3), and (4), in the form given. The reader can prove all six formulas using arbitrary limits as an exercise.

To prove (2), first note that the left side can be written

$$(x_1+y_1)+(x_2+y_2)+\ldots+(x_n+y_n) = (x_1+x_2+\ldots+x_n)+(y_1+y_2+\ldots+y_n)$$

$$= \sum_{i=1}^{n} x_i + \sum_{i=1}^{n} y_i = \text{right side of (2),}$$

because addition is both associative and commutative, i.e., numbers have the properties

$$(a+b)+c = a+(b+c) \quad \text{(associativity)}$$

and

$$a+b = b+a \quad \text{(commutativity).}$$

The distributivity of multiplication with respect to addition can be used to prove (3):

$$\sum_{i=1}^{n} cx_i = cx_1+cx_2+ \ldots +cx_n = c(x_1+ \ldots +x_n)$$

$$= c \sum_{i=1}^{n} x_i.$$

To prove (4), consider $x_i = a$ for $i = 1, 2, \ldots, n$. Then

$$\sum_{i=1}^{n} a = \sum_{i=1}^{n} x_i = (x_1+x_2+ \ldots +x_n) = (a+a+ \ldots +a)$$

$$= na.$$

Expressions (2)–(6) can be used to prove various results concerning the summation conventions:

(7) $$\frac{1}{n}\sum_{k=1}^{n} (x_k-\bar{x})^2 = \left(\frac{1}{n}\sum_{k=1}^{n} x_k^2\right) - \bar{x}^2$$

where $\bar{x}$ is expressed in (1). To show this, note that the left side of (7) can be written

(8) $$\frac{1}{n}\sum_{k=1}^{n} (x_k^2-2x_k\bar{x}+\bar{x}^2) = \frac{1}{n}\left(\sum_{k=1}^{n} x_k^2-2\bar{x}\sum_{k=1}^{n} x_k + \sum_{k=1}^{n} \bar{x}^2\right)$$

using (2) and (3). Expression (4) and the definition of $\bar{x}$ can be used to reduce the right side of (8) to

$$\frac{1}{n}\left\{\left(\sum_{k=1}^{n} x_k^2\right) - 2n(\bar{x})^2 + n(\bar{x})^2\right\} = \left(\frac{1}{n}\sum_{k=1}^{n} x_k^2\right) - \bar{x}^2.$$

Thus we have shown that the left side of (7) equals the right, and so (7) is valid.

(9) $\quad \displaystyle\sum_{k=1}^{n} k = \frac{n(n+1)}{2}.$

To prove (9), note that

$$\sum_{k=1}^{n} k = 1 + 2 + 3 + \ldots + (n-1) + n$$

$$= n + (n-1) + (n-2) + \ldots + 2 + 1$$

(changing the order of the terms of the sum)

$$= (n+1-1) + (n+1-2) + (n+1-3) + \ldots$$

$$+ [n+1-(n-1)] + (n+1-n)$$

$$= \sum_{k=1}^{n} (n+1-k),$$

and so

$$0 = \left(\sum_{k=1}^{n} k\right) - \sum_{k=1}^{n} [(n+1) - k]$$

$$= \left(\sum_{k=1}^{n} k\right) - \left(\sum_{k=1}^{n} (n+1)\right) + \sum_{k=1}^{n} k$$

$$= 2\left(\sum_{k=1}^{n} k\right) - n(n+1).$$

Therefore

$$\sum_{k=1}^{n} k = \frac{n(n+1)}{2}.$$

## EXERCISES

1 Prove expressions (5) and (6) by writing out the expressions on the left of the equal sign and then showing that this expanded expression can be rearranged so as to equal the expression on the right of the equal sign in each case.

2 Prove the following generalizations of (2)–(6):

(a) $\sum_{i=p}^{q} (x_i + y_i) = \sum_{i=p}^{q} x_i + \sum_{i=p}^{q} y_i,$

(b) $\sum_{i=p}^{q} c x_i = c \sum_{i=p}^{q} x_i,$

(c) $\sum_{i=p}^{q} a = (q-p+1)a,$

(d) $\sum_{i=p}^{q} \left( \sum_{j=r}^{s} a_{ij} \right) = \sum_{j=r}^{s} \left( \sum_{i=p}^{q} a_{ij} \right),$

(e) $\left( \sum_{i=p}^{q} a_i \right) \left( \sum_{j=r}^{s} b_j \right) = \sum_{i=p}^{q} \sum_{j=r}^{s} a_i b_j = \sum_{j=r}^{s} \sum_{i=p}^{q} a_i b_j,$

where $p \leq q$ and $r \leq s$.

3 Prove that

(a) $n \sum_{i=1}^{n} a_i^2 \geq \left( \sum_{i=1}^{n} a_i \right)^2$ for arbitrary $n$;

(b) $\sum_{i=1}^{n} \sum_{j=1}^{n} ij = \left[ \frac{n(n+1)}{2} \right]^2$;

(c) for $n < m < q$, $\sum_{i=n}^{q} x_i = \sum_{i=n}^{m} x_i + \sum_{i=m+1}^{q} x_i.$

4 If $f(x)$ is a function of $x$ and if $\{x_1, x_2, ..., x_n\}$ is a set of values that $x$ can assume, then the arithmetic mean of $f(x_1), ..., f(x_n)$ is

$$\bar{f}(x) = \frac{1}{n} \sum_{i=1}^{n} f(x_i).$$

Show that

(a) $\overline{(ax+b)} = a\bar{x}+b$, where $a$ and $b$ are constants,

(b) $\overline{(x-\bar{x})^2} = \overline{(x^2)} - (\bar{x})^2.$

5 If $\bar{y} = \frac{1}{n} \sum_{i=1}^{n} y_i$ and $\bar{x} = \frac{1}{n} \sum_{i=1}^{n} x_i$, then show that

$$\frac{1}{n} \sum_{i=1}^{n} (x_i - \bar{x})(y_i - \bar{y}) = \left( \frac{1}{n} \sum_{i=1}^{n} x_i y_i \right) - \bar{x}\bar{y}.$$

6 We also have an abbreviation for unions and intersections that is analogous to the $\sum$-notation for addition. If $A_1, A_2, ..., A_n$ are sets, then

$$A_1 \cup A_2 \cup A_3 \cup ... \cup A_n = \bigcup_{k=1}^{n} A_k$$

and

$$A_1 \cap A_2 \cap A_3 \cap ... \cap A_n = \bigcap_{k=1}^{n} A_k,$$

which are read 'the union (or intersection) of $A_k$ for $k = 1, 2, ..., n$.' Show, using conditions B1–B5 in section 0.1, that

(a) $A \cap \bigcup_{k=1}^{n} A_k = \bigcup_{k=1}^{n} (A \cap A_k),$

(b) $A \cup \bigcap_{k=1}^{n} A_k = \bigcap_{k=1}^{n} (A \cup A_k),$

(c) $\bigcup_{k=1}^{n} \left( \bigcup_{j=1}^{m} A_{kj} \right) = \bigcup_{j=1}^{m} \left( \bigcup_{k=1}^{n} A_{kj} \right).$

## 1.3
MEASURES OF DISPERSION

In section 1.1 we discussed the arithmetic mean, which could be described as the centre of gravity of a set of numerical values. It is often useful to know how concentrated the various numerical values are about this centre of gravity. Taking the other point of view, we might consider a measure of the dispersion of the values about their arithmetic mean. Then a small value of this measure of dispersion is indicative of high concentration.

Of the various measures of dispersion used, the simplest is the *range*[3] of the set of values $\{x_1, ..., x_n\}$:

(1) $R = \max(x_1, ..., x_n) - \min(x_1, ..., x_n)$

where $\max(x_1, ..., x_n)$ is the largest of the numerical values $x_1, ..., x_n$, and $\min(x_1, ..., x_n)$ is the smallest. Thus, for example, if $x_1 = 1$, $x_2 = 2$, $x_3 = 1$, $x_4 = 3$, $x_5 = 3$, $x_6 = 2$, then

$$\max(x_1, x_2, x_3, x_4, x_5, x_6) = 3,$$

$$\min(x_1, ..., x_6) = 1,$$

and the range is $R = 3 - 1 = 2$.

---

3  Not to be confused with the range of a function $f$, which is often defined to be the set of values $f(x)$ assigned by $f$ to $x$'s in the domain of $f$.

Because the range uses only two of the numerical values of the observations $x_1, ..., x_n$, it is not very sensitive as a description of the total collection.

A more sensitive measure of dispersion is the *mean absolute deviation* from the arithmetic mean:

$$(2) \quad d = \frac{1}{n} \sum_{i=1}^{n} |x_i - \bar{x}|$$

where the observations are of course $x_1, ..., x_n$. Remember that the vertical bars signify absolute value. Thus $|-1| = 1 = |1|$ for example.

EXAMPLE 1  If we take the observations from the *Tao Te Ching* considered in example 1 of section 1.1, i.e., if we take

$$x_1 = 29, \quad x_2 = 36, \quad x_3 = 36, \quad x_4 = 34, \quad x_5 = 31,$$

then

$$R = 34 - 29 = 5$$

and

$$d = \tfrac{1}{5} \sum_{i=1}^{5} |x_i - 33.5|$$

$$= \tfrac{1}{5}(4.2 + 2.8 + 2.8 + 0.8 + 2.2) = 2.56.$$

The reader will observe that in a situation where one or two of a set of observations lie at quite a distance from the arithmetic mean of the collection and all the rest of the observations in the collection lie near the mean, the mean absolute deviation gives a more realistic measure of the sample concentration than the range.

Another measure of dispersion is the *sample variance* of the observations $x_1, ..., x_n$:

$$(3) \quad s^2 = \frac{1}{n} \sum_{i=1}^{n} (x_i - \bar{x})^2.$$

For technical reasons to become clear later, the sample variance sometimes contains $n-1$ in the denominator instead of $n$. For the present, however, we shall use expression (3) as our definition.

For our observations from the *Tao Te Ching*,

$$s^2 = \tfrac{1}{5}[(4.2)^2 + (2.8)^2 + (2.8)^2 + (0.8)^2 + (2.2)^2]$$

$$= \tfrac{1}{5}[17.64 + 7.84 + 7.84 + 0.64 + 4.84] = 7.76.$$

It is obvious that the more concentrated the sample is about its arithmetic mean the smaller each of the indices (1), (2), and (3) will be.

It is easy to show that for non-negative real numbers $a_1, a_2, \ldots, a_n$

$$\sum_{i=1}^{n} a_i^2 \leq \left( \sum_{i=1}^{n} a_i \right)^2.$$

Thus in particular

$$\sum_{i=1}^{n} (x_i - \bar{x})^2 \leq \left( \sum_{i=1}^{n} |x_i - \bar{x}| \right)^2,$$

and so

$$\frac{1}{n^2} \sum_{i=1}^{n} (x_i - \bar{x})^2 \leq \left( \frac{1}{n} \sum_{i=1}^{n} |x_i - \bar{x}| \right)^2.$$

Thus the dispersion and the variance are related as follows:

(4) $s^2/n \leq d^2.$

It is often advantageous to use the square root of the sample variance or (sample) *standard deviation*

$$s = \sqrt{\frac{\sum\limits_{i=1}^{n} (x_i - \bar{x})^2}{n}}$$

to measure dispersion because it is in the same units as the variate. Thus in terms of the standard deviation (4) becomes

(5) $s \leq \sqrt{nd},$

when we multiply both of its sides by $n$ and take the square root.

Using the methods of section 1.2, we can easily show that

(6) $s^2 = \dfrac{1}{n} \left( \sum\limits_{i=1}^{n} x_i^2 \right) - (\bar{x})^2,$

and if $y_i = ax_i + b$ and $s_y^2$ and $s_x^2$ are respectively the variances of the observations $y_1, \ldots, y_n$ and $x_1, \ldots, x_n$, then

$$s_y^2 = \frac{1}{n} \sum_{i=1}^{n} (y_i - \bar{y})^2 = \frac{1}{n} \sum_{i=1}^{n} (ax_i + b - a\bar{x} - b)^2$$

$$= \frac{1}{n} \sum_{i=1}^{n} a^2 (x_i - \bar{x})^2 = \frac{a^2}{n} \sum_{i=1}^{n} (x_i - \bar{x})^2 = a^2 s_x^2.$$

This can be put more succinctly in the form

(7) $s_{ax+b}^2 = a^2 s_x^2.$

EXERCISES

1 Take your results in exercises 1 and 3 of section 1.1 and compute $R$, $d$, and $s$ for each of the samples obtained in exercise 1(a), (b), and (c).

2 Find $s$ for the sample given in example 1 of this section. (If you have forgotten how to compute the square root and have no computing machine available, you can sometimes use a mathematical table of square roots. These usually give the square roots of integers between 1 and 100 or 1 and 1000. In this case we are lucky because

$$s^2 = 10.25 = 1025 \times 10^{-2} = 25 \times 41 \times 10^{-2}$$

so $s = \sqrt{25} \times \sqrt{41} \times \sqrt{10^{-2}} = \tfrac{1}{2}\sqrt{41}$.)

3 If $X_1$, $X_2$, ..., $X_n$ are $n$ variates and another variate $Y$ is constructed from these by the formula

$$Y = \sum_{i=1}^{n} a_i X_i$$

where the $a_i$ are constants, and if a sample $(x_{1i}, x_{2i}, ..., x_{mi})$ composed of $m$ observations from each of the $X_i$ ($i = 1, 2, 3, ..., n$) is obtained, then show that the sample mean and variance of $Y$ take the forms

$$\bar{Y} = \sum_{i=1}^{n} a_i \bar{X}_i \quad \text{and} \quad s_Y{}^2 = \sum_{i=1}^{n} a_i{}^2 s_{X_i}{}^2$$

where $\bar{X}_i$ is the sample mean of $(x_{1i}, x_{2i}, ..., x_{mi})$ and $s_{X_i}{}^2$ its sample variance.

## 1.4
### CURVE FITTING

The *Tao Te Ching* contains 81 chapters. Suppose we select at random seven of these, say $C_1$, $C_2$, ..., $C_7$, and count $x_i$, the length of chapter $C_i$, and $y_i$, the number of different Chinese characters used in $C_i$. Using some random device, we select seven numbers from 1 to 81. For example, we might obtain

34, 11, 50, 18, 48, 75, 64.

TABLE 1.1

| Chapter | $x$ | $y$ |
|---------|-----|-----|
| 34 | 22 | 20 |
| 11 | 49 | 24 |
| 50 | 80 | 42 |
| 18 | 26 | 22 |
| 48 | 40 | 23 |
| 75 | 54 | 26 |
| 64 | 91 | 55 |

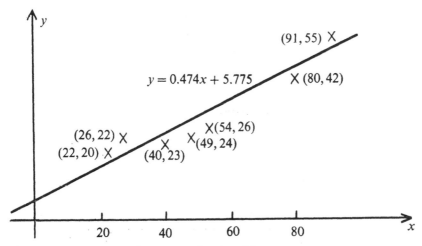

FIGURE 1.2   Least squares linear approximation of data.

Table 1.1 gives the results for these chapters. In figure 1.2 the ordered pairs $(x, y)$ are plotted. Since the values of $y$ that are given here do not appear to fit any pattern, it would not be profitable to obtain the equation of, say, a polynomial curve

$$y = \sum_{i=0}^{n} a_i x^i$$

which passes exactly through these seven points. However, it may be useful to obtain the equation of the straight line which best approximates the data we have obtained (figure 1.2).

For a measure of the degree of approximation, it is theoretically most advantageous to use the sum of the squares of the vertical distances of the data-points from the approximating straight line. Thus if $(x_1, y_1), ..., (x_n, y_n)$ are data-points and the approximating straight line has the equation

(1)   $y = ax + b$,

then in each case the vertical distance of $(x_i, y_i)$ from the straight line of equation (1) is

$$d_i = |y_i - (ax_i + b)|,$$

as in figure 1.3. Thus the sum of the squares of these distances is

(2)   $D = \sum_{i=1}^{n} d_i^2 = \sum_{i=1}^{n} [y_i - (ax_i + b)]^2$.

Our problem is then to find the $a$ and $b$ that make $D$ the smallest possible. The values of $a$ and $b$ which render $D$ smallest are sometimes called the *linear*

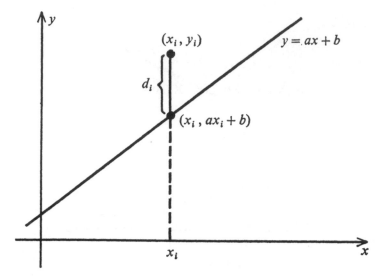

*regression coefficients*, and the line with the equation $y = ax+b$ is sometimes called the *least squares linear approximation* to the given data.

This problem can be solved using differential calculus, and we shall proceed to solve it. Readers whose knowledge of this subject is limited, or who are not interested in the theoretical development, may omit the next few paragraphs and rejoin the discussion at expression (8).

It is clear that $D$ is a function of $a$ and $b$ simultaneously, i.e.,

$$D = D(a, b).$$

If it is to have a low point or minimal value, this value will occur only when

(3) $\quad \dfrac{\partial D}{\partial a} = \dfrac{\partial D}{\partial b} = 0.$

From (2) we obtain

(4) $\quad \dfrac{\partial D}{\partial a} = -2 \sum\limits_{i=1}^{n} [y_i - (ax_i + b)]x_i = 0$

and

(5) $\quad \dfrac{\partial D}{\partial b} = -2 \sum\limits_{i=1}^{n} [y_i - (ax_i + b)] = 0.$

Thus at the minimum point (5) yields

$$\sum_{i=1}^{n} y_i - a \left( \sum_{i=1}^{n} x_i \right) - nb = 0$$

or better

(6) $\bar{y} - a\bar{x} - b = 0,$

and (4) yields

(7) $\dfrac{1}{n}\left(\displaystyle\sum_{i=1}^{n} y_i x_i\right) - \dfrac{a}{n}\left(\displaystyle\sum_{i=1}^{n} x_i^2\right) - b\bar{x} = 0.$

Equations (6) and (7) taken together yield the results

(8) $a = \dfrac{\left(\dfrac{1}{n}\displaystyle\sum_{i=1}^{n} x_i y_i\right) - \bar{x}\bar{y}}{\dfrac{1}{n}\left(\displaystyle\sum_{i=1}^{n} x_i\right)^2 - \bar{x}^2} = \dfrac{\dfrac{1}{n}\displaystyle\sum_{i=1}^{n}(x_i - \bar{x})(y_i - \bar{y})}{s_x^2}$

and

(9) $b = \bar{y} - \dfrac{\bar{x}}{s_x^2}\left(\dfrac{1}{n}\displaystyle\sum_{i=1}^{n}(x_i - \bar{x})(y_i - \bar{y})\right).$

The expression

(10) $r(x, y) = \dfrac{1}{n}\displaystyle\sum_{i=1}^{n}(x_i - \bar{x})(y_i - \bar{y})$

is sometimes called the *sample covariance* of the variates $x$ and $y$.

Using our seven observations from the *Tao Te Ching*, we obtain the results displayed in table 1.2. Thus

$$\bar{x} = 51.714, \quad \bar{y} = 30.286, \quad a = 0.474, \quad b = 5.775.$$

In figure 1.2, we have drawn the line with equation

$$y = ax + b = (0.474)x + 5.775.$$

Because of the fact that each writer has a finite writing vocabulary $N$, it should be clear that the variate $y$ is always bounded above by $N$, i.e., $y \leq N$, no matter how large $x$ might be. Therefore a straight line will not necessarily fit this sort of data when the range of possible values of $x$ is large. Indeed, for all values of $x$, the points $(x, y)$ lie below the horizontal line

$$\{(x, y) \mid y = N\}.$$

However, it is still of interest to fit a straight line to data in which the values of $x$ are restricted to a relatively short interval.

Where the relationship between two variates $x$ and $y$ is involved, the sample covariance is a measure which is sensitive to the linear dependence of the $x_i$ on

TABLE 1.2

| $x_i$ | $y_i$ | $x_i - \bar{x}$ | $y_i - \bar{y}$ | $(x_i - \bar{x})^2$ | $(y_i - \bar{y})^2$ | $(x_i - \bar{x})(y_i - \bar{y})$ | $d_i = y_i - (ax_i + b)$ | $d_i^2$ |
|---|---|---|---|---|---|---|---|---|
| 22 | 20 | −29.71 | −10.29 | 882.68 | 105.88 | 305.72 | 3.80 | 14.44 |
| 49 | 24 | −2.71 | −6.29 | 7.34 | 39.56 | 17.05 | −5.00 | 25.00 |
| 80 | 42 | 28.29 | 11.71 | 880.32 | 137.12 | 331.28 | −1.70 | 2.89 |
| 26 | 22 | −25.71 | −8.29 | 661.00 | 68.72 | 213.14 | 3.90 | 15.21 |
| 40 | 23 | −11.71 | −7.29 | 137.12 | 53.14 | 85.37 | −1.74 | 3.03 |
| 54 | 26 | 2.29 | −4.29 | 5.24 | 18.40 | −9.82 | −5.37 | 28.84 |
| 91 | 55 | 39.29 | 24.71 | 1543.70 | 610.59 | 970.86 | 6.09 | 37.09 |
| TOTAL 362 | 212 | | | 4037.43 | 1033.41 | 1913.57 | | 126.50 |

the $y_i$. It is more interesting to consider the covariance normalized by dividing by $s_x s_y$. This quantity,

$$(11) \quad \rho(x, y) = \frac{r(x, y)}{s_x s_y},$$

is called the *sample correlation coefficient*. It has the advantage of lying at all times between $-1$ and $1$. The reader can see that if all the points of the sample are of the form $(a_i, a_i)$, i.e., if they all lie on the line with equation $y = x$, then

$$r(x, y) = \frac{1}{n}\left(\sum_{i=1}^{n} a_i^2\right) - \left(\frac{1}{n}\sum_{i=1}^{n} a_i\right)^2$$

and

$$s_x = s_y = \sqrt{\frac{1}{n}\left(\sum_{i=1}^{n} a_i^2\right) - \left(\frac{1}{n}\sum_{i=1}^{n} a_i\right)^2}.$$

Thus for such a sample $\rho(x, y) = 1$. If, however, the sample were composed of points $(a_i, -a_i)$, then it is easy to see that

$$\rho(x, y) = -1.$$

Suppose in a third case the sample took the form

$$(-1, 1), \ (1, -1), \ (-1, -1), \ (1, 1).$$

Then $r(x, y) = 0$ and so $\rho(x, y) = 0$.

In general if the sample pairs $(x_i, y_i)$ group themselves about a line $y = ax + b$ where $a$, the slope of the line, is positive, the sample correlation coefficient will be near one; if they group themselves about a line with slope $a < 0$, $\rho(x, y)$ will be near $-1$; and if they scatter in some random fashion, then $\rho(x, y)$ will be near zero.

For the *Tao Te Ching* sample, table 1.2 yields a sample correlation coefficient

$$\rho(x, y) = \frac{\frac{1}{n}\sum (x_i - \bar{x})(y_i - \bar{y})}{\sqrt{\frac{1}{n}\sum (x_i - \bar{x})^2}\sqrt{\frac{1}{n}\sum (y_i - \bar{y})^2}}$$

$$= \frac{\sum (x_i - \bar{x})(y_i - \bar{y})}{\sqrt{\sum (x_i - \bar{x})^2 \cdot \sum (y_i - \bar{y})^2}}$$

$$= \frac{1913.57}{\sqrt{4037.43}\sqrt{1033.41}} = \frac{1913.57}{(63.54)(32.15)} = 0.937,$$

which indicates a fairly good fit.

In general, other functions than $ax+b$ can be used to approximate bivariate data. For example, a quadratic function

$$y = ax^2+bx+c$$

might be employed; in that case we would try to minimize the function

$$D = \sum_{i=1}^{n} [y_i-(ax_i^2+bx_i+c)]^2$$

by setting

$$\frac{\partial D}{\partial a} = \frac{\partial D}{\partial b} = \frac{\partial D}{\partial c} = 0$$

and trying to obtain expressions for $a$, $b$, and $c$ in terms of the $(x_i, y_i)$'s.

The reader may have observed that the computation involved in obtaining even the coefficients of linear regression for a small sample like the one we have just taken from the *Tao Te Ching* is somewhat weighty, and he might be forgiven for feeling that linear regression for a larger sample and quadratic regression in general could give rise to computation beyond his endurance. Fortunately there is the electronic computer. Most computer installations and even standard desk model electronic computers have packaged programs for least squares approximations to various classes of functions. All the researcher needs to do is enter his data appropriately and the packaged program will do the rest for him.

In 29 paragraphs chosen at random from *La Barraca*, a novel by Vicente Blasco Ibáñez, the following pairs $(x, y)$ were obtained, where $x$ is the number of words in the paragraph and $y$ the number of different words in the paragraph:

(35, 29), (50, 39), (40, 34), (62, 48), (43, 33), (79, 57), (101, 67), (46, 39), (44, 38), (62, 48), (37, 33), (63, 50), (172, 107), (221, 134), (32, 29), (70, 47), (39, 32), (35, 29), (140, 100), (32, 27), (22, 20), (103, 73), (32, 26), (42, 37), (69, 50), (60, 49), (99, 68), (76, 61), (6, 6).

Using one of the packages mentioned in the previous paragraph, the author obtained the following values for $a$, $b$, and $\rho$ for a linear regression analysis fitting the line $y = ax+b$:

(12) $a = 0.587,\quad b = 9.919,\quad \rho = 0.993.$

Using a second package-program which gives $a_0, a_1, a_2$ such that

$$y = a_2x^2+a_1x+a_0$$

best fits the data in the least squares sense, we obtain

$$a_2 = -0.0007,\quad a_1 = 0.736,\quad a_0 = 4.603.$$

In order to see how much of an improvement the quadratic approximation is over the linear, we can compute the *residual sum of squares*

$$D_l = \sum_{i=1}^{n} [y_i - (ax_i + b)]^2$$

in the linear case and

$$D_q = \sum_{i=1}^{n} [y_i - (a_2 x_i^2 + a_1 x_i + a_0)]^2$$

in the quadratic case. For our data

$$D_l = 310.80 \quad \text{and} \quad D_q = 191.32,$$

the latter being rather a better fit.

EXERCISES

1 From the work which you used for exercise 1 in section 1.1 select 25 paragraphs at random and plot for each the point $(x, y)$ where $x$ is the length in words of the paragraph and $y$ is the type-count of the paragraph.

2 (a) Find $a$ and $b$ such that $y = ax + b$ is the best-fitting straight line for your data. (This may require the use of a desk calculator.)
   (b) Find $\rho(x, y)$ for your data.

3 In your sample, can you hypothesize what factors might contribute to lowering the type-to-token ratio? Is the subject matter of the passage more important than its grammatical structure in this regard?

4 Suppose in a series of $n$ samples we obtain as sample values of $(x, y)$

$$(x_1, ax_1 + b), (x_2, ax_2 + b), ..., (x_n, ax_n + b).$$

What is the value of $\rho(x, y)$ for this sample?

5 Take a die and roll it twice, recording the result of the first roll as $x$ and that of the second as $y$. Perform this experiment five times to obtain

$$(x_1, y_1), ..., (x_5, y_5).$$

Use these pairs to compute $\rho(x, y)$. Do you expect that in general $\rho$ will be near zero? If so, why?

6 Show that $D$ given by (2) can be written $D = n(1 - \rho^2) s_y^2$.

7 Given the data (1, 1), (2, 1), (3, 3), find $a$, $b$, $c$ such that

$$y = ax^2 + bx + c$$

best fits the data. Can you explain why the result comes out as it does?

8 Show that if $y = ax+b$ is the equation of the line that yields a minimum $D$ for the sample points $(x_1, y_1), ..., (x_n, y_n)$, then

$$\bar{y} = a\bar{x}+b.$$

9 (a) Plot on a graph the points corresponding to the paragraphs taken at random from *La Barraca* and draw the linear regression line

$$y = ax+b$$

where $a$ and $b$ are given in expression (12).
   (b) Plot the parabola

$$y = -0.0007x^2+0.736x+4.603$$

on the same sheet of paper in order to ascertain the difference between the two curves.


## 1.5

### THE HISTOGRAM

Suppose we have obtained a sample $(x_1, ..., x_n)$ of a variate $X$. If $n$ is large, the distribution of the numbers $x_i$ along the real-number line is not necessarily evident from the list $x_1, x_2, ..., x_n$ of outcomes in the sample. The following procedure can be used to obtain a quite satisfactory picture of this distribution.

All the outcomes $x_1, ..., x_n$ in the sample will lie between two numbers $a$ and $b$, i.e., $a$ and $b$ can be found such that for all $i$

$$a < x_i < b.$$

Suppose the interval from $a$ to $b$ (figure 1.4) is divided into $m$ equal subintervals, or *subclasses*,

$$I_k = \{x \mid a_{k-1} < x \leq a_k\}$$

where the points $a_0, a_1, ..., a_m$, sometimes called *class limits*, are given by the equation

$$a_k = a + \frac{k(b-a)}{m}.$$

Each $x_i$ in the sample will lie in exactly one of the subintervals $I_k$. Let $f_k$ be the number of outcomes $x_i$ that lie in $I_k$. For each $k$ construct a rectangle of height $f_k$ and base $(b-a)/m$ upon the interval $I_k$. The resulting assemblage of rectangles is called a *histogram*.

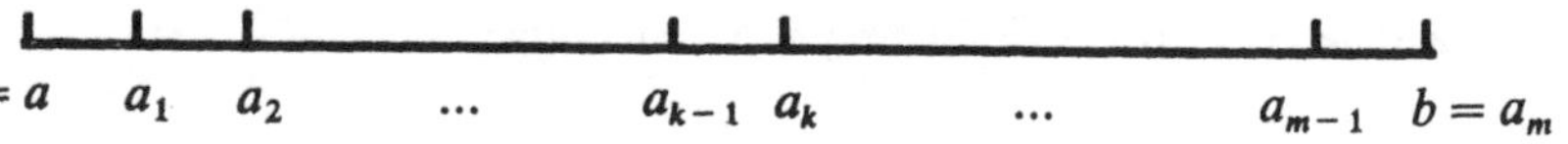

FIGURE 1.4

TABLE 1.3

| $k$ | No. of $x_i$ in $I_k = f_k$ |
|---|---|
| 1 | 1 |
| 2 | 3 |
| 3 | 4 |
| 4 | 6 |
| 5 | 7 |
| 6 | 8 |
| 7 | 7 |
| 8 | 5 |
| 9 | 0 |
| 10 | 2 |

As a simple example suppose that a sample $(x_1, ..., x_{43})$ is obtained and all the values lie between 0 and 10. We might divide that interval into subintervals $I_1, ..., I_{10}$ each of length one by choosing

$$a_0 = 0, \quad a_1 = 1, \quad ..., \quad a_9 = 9, \quad a_{10} = 10.$$

Suppose that the $x_i$ distribute themselves among the subintervals

$$I_k = \{x \mid k-1 < x \leqq k\}$$

for $k = 1, 2, ..., 10$ as indicated in table 1.3. The resulting histogram is obtained by erecting 10 rectangles each with base the interval $I_k$ and height $f_k$ as indicated in figure 1.5.

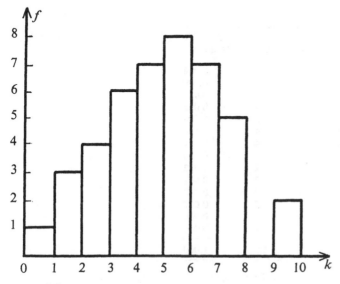

FIGURE 1.5

To see how a histogram shows the nature of the distribution of sample values for different choices of interval length, consider the following concrete example.

EXAMPLE 1   In a sample of 28 fifty-word passages chosen at random from the novel *Riders in the Chariot* by the Australian writer Patrick White, we obtain the following numbers of different words (i.e. word types) in each of the passages:

40, 40, 43, 40, 44, 43, 33, 36, 35, 40, 41, 38, 39, 39, 37, 38, 42, 37, 38, 40, 42, 41, 38, 39, 39, 44, 37, 46.

The selection of the passages was carried out by first noting that there were 532 pages in the Viking edition of the book, then selecting 28 of these pages at random using a table of random numbers, and finally selecting a line from each of these pages at random: there are 39 lines on each page and one of these was chosen using the random-number table again.

To obtain a clear picture of the distribution of the sample, we make a histogram as follows. First make a tally chart as in table 1.4. A histogram corresponding to the sample and the class limits of table 1.4 is easily constructed and appears here as in figure 1.6.

If we choose the class limits differently, say as

32.5, 34.5, 36.5, ..., 46.5,

the histogram in figure 1.6a is obtained.

TABLE 1.4

| Class limits[4] | Tally mark | Frequency | Relative frequency |
|---|---|---|---|
| 32.5–33.5 | . | 1 | 0.036 |
| 33.5–34.5 | | | 0.000 |
| 34.5–35.5 | . | 1 | 0.036 |
| 35.5–36.5 | . | 1 | 0.036 |
| 36.5–37.5 | : . | 3 | 0.108 |
| 37.5–38.5 | : : | 4 | 0.144 |
| 38.5–39.5 | : : | 4 | 0.144 |
| 39.5–40.5 | :⁄: | 5 | 0.180 |
| 40.5–41.5 | . . | 2 | 0.072 |
| 41.5–42.5 | . . | 2 | 0.072 |
| 42.5–43.5 | . . | 2 | 0.072 |
| 43.5–44.5 | . . | 2 | 0.072 |
| 44.5–45.5 | | | 0.000 |
| 45.5–46.5 | . | 1 | 0.036 |
| TOTAL | | 28 | 1.000 |

4   With integer-valued data it is standard practice to choose class limits so that they fall mid-way between integers.

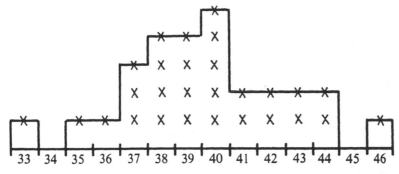

FIGURE 1.6

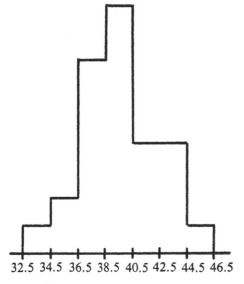

FIGURE 1.6a

Note that the median of this sample (see exercise 5 in section 1.1) is 39, the mean is $\bar{X} = 39.61$, and the variance is $s_x^2 = 7.94$.

There are samples where the central tendency of the distribution of values may not be so clear cut. For example, in sampling *Human Knowledge: Its Scope and Limits* (by Bertrand Russell) for the number of articles (*a, an, the*) in 50-word passages, we obtained the histogram in figure 1.7 from a random sample of 30 such passages chosen from the text. Here it is clear that 3 and 8 are the most favoured numbers of articles, but it is not clear that this represents the state of affairs for Russell's writing in *Human Knowledge* as a whole. Will a larger sample reveal that Russell distributes his articles in such a way that 5-article 50-word

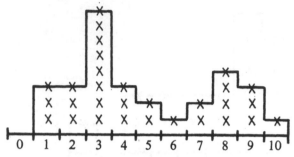

FIGURE 1.7 Histogram of the number of articles in 50-word passages from *Human Knowledge* by Bertrand Russell. Each cross stands for a 50-word passage and is entered in the column corresponding to the number of articles in that passage.

passages tend to occur seldom while 8- and 3-article passages are the norm? Intuitively, this seems highly unlikely; doubtless a larger sample will tend to fill in the columns between 3 and 8 revealing a *mode* of perhaps 4 or 5 articles in a 50-word passage. The *mode* of a histogram is that sample value that occurs

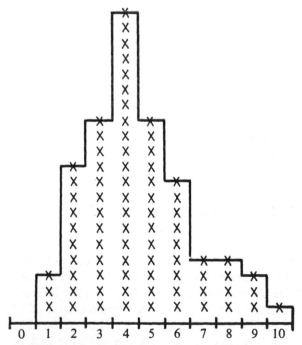

FIGURE 1.8 Histogram of the number of articles in each of eighty 50-word passages selected at random (including the 30 passages of figure 1.7) from *Human Knowledge*.

most often. In figure 1.7 it is 3. Usually we expect a 'nice' histogram to rise to a peak and then fall away without displaying the jagged appearance that occurs in figure 1.7. If we increase the sample size to 80 randomly selected 50-word passages, we obtain the histogram of figure 1.8. Note that the central tendency has asserted itself. The mode of the histogram is 4, its mean is 4.475, and its median is 4.

EXERCISES

1 Suppose that a sample $(x_1, ..., x_n)$ of a variate $X$ is obtained and that all the $x_i$ lie between $a$ and $b$.

(a) A histogram is constructed for this sample using intervals of length $(b-a)/m$. Show that each $x_i$ in

$$I_k = \left\{ x \left| a + \frac{(k-1)(b-a)}{m} < x \leq a + \frac{k(b-a)}{m} \right. \right\}$$

differs from

$$a + \frac{(k-\frac{1}{2})(b-a)}{m}$$

by a number which is less than or equal to $(b-a)/2m$, i.e., if $x_i \in I_k$, then

$$\left| x_i - \left( a + \frac{(k-\frac{1}{2})(b-a)}{m} \right) \right| \leq \frac{b-a}{2m}.$$

(b) Then show that

$$\left| \bar{X} - \frac{1}{n} \sum_{i=1}^{m} \left( a + \frac{(k-\frac{1}{2})(b-a)}{m} \right) f_i \right| \leq \frac{1}{n} \sum_{i=1}^{m} \frac{b-a}{2m} f_i,$$

and hence that the difference between the sample mean $\bar{X}$ and the approximation obtained from the histogram is less than $(b-a)/2m$. Thus the mean of a sample can be obtained (approximately) directly from a histogram of the sample.

2 In the sample taken from *Riders in the Chariot* (see example 1) the common-noun counts in the 28 passages were as follows:

6, 9, 8, 11, 8, 10, 9, 6, 10, 7, 7, 9, 8, 9, 8, 9, 6, 6, 10, 10, 6, 7, 10, 8, 9, 8, 11, 8.

(a) Make a histogram of this sample.
(b) Find its mean and standard deviation.

3 In the sample from *Riders in the Chariot*, for each 50-word passage the number of verbs $v$ and adjectives $a$ was counted and the following pairs $(v, a)$ were obtained:

(8, 8), (9, 4), (7, 3), (7, 4), (6, 7), (10, 3), (5, 5), (9, 4), (7, 1), (8, 5), (5, 6), (7, 5), (7, 3), (10, 2), (5, 7), (8, 4), (6, 6), (6, 5), (4, 1), (6, 4), (6, 7), (5, 4), (7, 6), (9, 3), (4, 3), (8, 5), (5, 7), (10, 4).

(a) In Doležel and Bailey (1969), Antosch (p. 57) suggests that

$$\text{VAR} = \frac{\text{no. of verbs}}{\text{no. of adjectives}} = \frac{v}{a}$$

be used as a measure of the active quality in writing. Compute VAR for each of the passages sampled.

(b) Make a histogram of VAR for this sample of 28 passages using class limits 0, 1, 2, ...

(c) Make a histogram of VAR using class limits 0, $\frac{1}{2}$, 1, $1\frac{1}{2}$, 2, ...

(d) Find the mean and standard deviation of VAR.

(e) Find the correlation between $v$ and $a$ for this sample.

4 Find the variance of the 80-passage sample from *Human Knowledge* using the information in figure 1.8.

## 1.6
### SAMPLE MOMENTS

Given a sample $(x_1, x_2, ..., x_n)$ of a variate $X$, we can construct the *arithmetic mean* $\mathcal{M}(g(X))$ of any function $g$ of $X$ for this sample. It is by definition

$$(1) \quad \mathcal{M}(g(X)) = \frac{1}{n} \sum_{i=1}^{n} g(x_i).$$

Examples of (1) that we have already discussed are

$$\bar{X} = \mathcal{M}(X) = \frac{1}{n} \sum_{i=1}^{n} x_i$$

when $g(x) = x$ and

$$s_X^2 = \mathcal{M}((X-\bar{X})^2) = \frac{1}{n} \sum_{i=1}^{n} (x_i-\bar{X})^2$$

when $g(x) = (x-\bar{X})^2$.

If $g(x) = (x-c)^r$ where $r = 1, 2, ...$, then

$$\mathcal{M}(g(X)) = \frac{1}{n} \sum_{i=1}^{n} (x_i-c)^r$$

is the *rth sample moment of X* (for the sample $(x_1, x_2, ..., x_n)$) *about c*. Thus the mean $\bar{X}$ is the first sample moment about $c = 0$ and $s_X^2$ is the second sample moment about the sample mean of $X$. In general, the moments about the origin and about the mean are of special importance. Thus for the sample $(x_1, x_2, ..., x_n)$ we define

(2) $\quad m_r = \mathcal{M}(X^r) = \dfrac{1}{n} \displaystyle\sum_{i=1}^{n} x_i^r$

as the *rth sample moment about the origin* (that is about the point $O$ on the axis containing the sample points) and

(3) $\quad \mu_r = \mathcal{M}((X - \bar{X})^r) = \dfrac{1}{n} \displaystyle\sum_{i=1}^{n} (x_i - \bar{X})^r$

as the *rth central sample moment,* that is the *r*th moment about the (sample) mean $\bar{X}$.

We have religiously included the word 'sample' before 'moment' because in section 3.2 we come to discuss the moments of a distribution. In that same chapter, we shall find that there is a useful relationship between these two kinds of moments. For the remainder of this section, however, we adopt the convention that 'moment' means 'sample moment' in its various forms as defined above.

As we have already observed, $\bar{X}$ locates the centre of gravity of the sample, and hence (approximately) of a histogram constructed from the sample, and $s_X^2$ measures the dispersion of the sample and hence (approximately) of its histogram.[5] The higher moments can be used to measure other properties of a histogram constructed from a given sample. In particular,

$\quad \beta_1 = \mu_3^2 / \mu_2^3$

is a measure of the *skewness* (asymmetry) of the sample (histogram). If the sample is symmetrically distributed about $\bar{X}$, then, and only then, $\beta_1 = 0$. If $\mu_3 < 0$ it is skewed to the left (figure 1.9), and if $\mu_3 > 0$ it is skewed to the right.

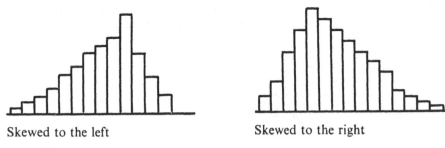

Skewed to the left          Skewed to the right

FIGURE 1.9

5  This is true only approximately in that the mean $\bar{X}$, for example, will be within one-half the interval-length of the histogram from the centre of gravity of the histogram. See exercise 1 in section 5.1. Of course if $X$ can assume only integral values and the histogram intervals are one unit in length, then the centres of gravity of the histogram and the sample are the same.

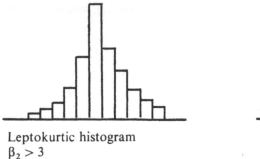 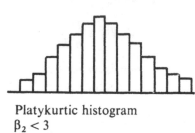

Leptokurtic histogram
$\beta_2 > 3$

Platykurtic histogram
$\beta_2 < 3$

FIGURE 1.10

The histogram on the left in figure 1.9 is skewed to the left ($\mu_3 < 0$) and the one on the right is skewed to the right ($\mu_3 > 0$).

A second quantity,

$$\beta_2 = \mu_4/\mu_2{}^2,$$

measures *kurtosis* (peakedness). If $\beta_2 < 3$ the histogram is said to be *platykurtic* (tending to be flat), and if $\beta_2 > 3$ it is *leptokurtic* (tending toward peakedness). The kurtosis value 3 is the peakedness expected of a sample whose distribution is 'normal.' In chapters 3 and 4 we consider the properties of these 'normal' samples in detail. In figure 1.10 we illustrate a platykurtic and a leptokurtic histogram.

Sometimes it is convenient when considering a variate with discrete values, such as the number of syllables in a word or the number of articles in a 50-word passage of prose, to render the sample in the form of a frequency count as in tables 1.5 and 1.6.

EXAMPLE 1 To obtain the mean of the sample in table 1.5, we total all the observations and divide by 1007. Of these observations 713 are one-syllable

TABLE 1.5

Syllable counts of the first 1007 words in *The Wapshot Chronicle* by John Cheever

| No. of syllables per word | No. of words observed |
|---|---|
| 1 | 713 |
| 2 | 221 |
| 3 | 54 |
| 4 | 15 |
| 5 | 4 |
| 6 | 0 |
| TOTAL | 1007 |

TABLE 1.6

Article (*a, an, the*) counts of eighty-three
50-word passages selected at random
from *The Wapshot Chronicle*

| No. of articles per passage | No. of passages observed |
|:---:|:---:|
| 0 | 8 |
| 1 | 8 |
| 2 | 12 |
| 3 | 11 |
| 4 | 14 |
| 5 | 4 |
| 6 | 8 |
| 7 | 8 |
| 8 | 6 |
| 9 | 2 |
| 10 | 2 |
| TOTAL | 83 |

words, 221 are two-syllable, etc. Thus

$$\bar{X} = \frac{1}{1007}(1 \times 713 + 2 \times 221 + \ldots + 4 \times 5 + 0 \times 6) = 1.387.$$

More generally, if a sample $(x_1, \ldots, x_n)$ of a variate $X$ contains $f_1$ observations of $a_1, f_2$ observations of $a_2, \ldots,$ and $f_m$ observations of $a_m$, then the sample mean is given by

$$\bar{X} = \sum_{i=1}^{m} f_i a_i \Big/ \sum_{i=1}^{m} f_i = \frac{1}{n} \sum_{i=1}^{m} f_i a_i,$$

and for the other moments

$$m_r = \frac{1}{n} \sum_{i=1}^{m} f_i a_i^r$$

and

$$\mu_r = \frac{1}{n} \sum_{i=1}^{m} f_i (a_i - \bar{X})^r.$$

Thus the sample variance of the sample of table 1.5 is

$$\mu_2 = s_X^2 = \frac{1}{1007} \{713(1-1.387)^2 + 221(2-1.387)^2 + \ldots + 0(6-1.387)^2\}$$

$$= 0.482.$$

To obtain measures of its skewness and kurtosis we must obtain

$$\mu_3 = \frac{1}{1007} \{713(1-1.387)^3 + 221(2-1.387)^3 + \ldots + 0(6-1.387)^3\}$$

$$= 0.688,$$

$$\mu_4 = \frac{1}{1007} \{713(1-1.387)^4 + 221(2-1.387)^4 + \ldots + 0(6-1.387)^4\}$$

$$= 1.781.$$

Then we obtain

$$\beta_1 = 4.23 \quad \text{and} \quad \beta_2 = 7.67.$$

Thus the distribution of sample values in table 1.5 is skewed to the right, as can be seen if we construct a histogram for the sample. Since $\beta_2$ is much greater than 3 the distribution is leptokurtic, i.e., peaked.

The moments about the origin and the central moments are not independent of one another. Indeed, as in the case of the variance where

$$\mu_2 = \frac{1}{n}\left(\sum_{i=1}^{n} x_i^2\right) - m_1^2 = m_2 - m_1^2,$$

we can express the central moments in terms of the moments about the origin. If we take $r = 3$ for example, we find by a simple but tedious computation that

$$\mu_3 = \frac{1}{n}\sum_{i=1}^{n}(x_i - m_1)^3 = \frac{1}{n}\sum_{i=1}^{n}(x_i^3 - 3x_i^2 m_1 + 3x_i m_1^2 - m_1^3)$$

$$= \frac{1}{n}\sum_{i=1}^{n} x_i^3 - \frac{3m_1}{n}\sum_{i=1}^{n} x_i^2 + \frac{3m_1^2}{n}\sum_{i=1}^{n} x_i - \frac{nm_1^3}{n}$$

$$= m_3 - 3m_2 m_1 + 3m_1^3 - m_1^3 = m_3 - 3m_2 m_1 + 2m_1^3.$$

In general it can be shown that for $r \geq 2$

$$(4) \quad \mu_r = \sum_{i=0}^{r}(-1)^i\binom{r}{i} m_{r-i} m_1^i$$

where the symbol $\binom{r}{i}$ stands for the number of different ways a subset containing $i$ elements can be selected from a set containing $r$ elements. For $0 \leq i \leq r$

$$\binom{r}{i} = \frac{r!}{i!\,(r-i)!}$$

where $0! = 1$ and, for $x > 0$,

$$x! = 1 \cdot 2 \cdot 3 \ldots x.$$

EXAMPLE 2 Expression (4) allows for a simplification in the computation of moments. For the data given in table 1.6, the computation needed to find the first four sample moments is greatly shortened using (4). We first find $m_1 = 3.99$, $m_2 = 22.95$, $m_3 = 154.20$, $m_4 = 1137.94$, and then the central moments can be computed using (4):

$$s_x^2 = \mu_2 = m_2 - m_1^2 = 7.03,$$

$$\mu_3 = m_3 - 3m_2 m_1 + 2m_1^3 = 6.53,$$

$$\mu_4 = m_4 - 4m_3 m_1 + 6m_2 m_1^2 - 3m_1^4 = 108.76.$$

Thus $\beta_1 = 0.123$, indicating a fairly symmetric distribution about the mean, and $\beta_2 = 2.161$, indicating a platykurtic distribution, i.e., with tendency toward flatness relative to a normal distribution.

## EXERCISES

1 Use the binomial theorem, which states that

$$(a+b)^n = \sum_{i=0}^{n} \binom{n}{i} a^{n-i} b^i,$$

to show that expression (4) holds.

2 Some exercises in algebra: show that

(a) $n(n-1) = \dfrac{n!}{(n-2)!}$,

(b) $\dfrac{n(n-1)}{2!} = \dfrac{n!}{(n-2)!\,2!} = \binom{n}{2}$,

(c) $\dfrac{n(n-1)(n-2)}{3\cdot 2} = \dfrac{n!}{(n-3)!\,3!} = \binom{n}{3}$,

(d) $\binom{n}{i} = \dfrac{n(n-1)...(n-i+1)}{i(i-1)...2\cdot 1}$,

(e) $\binom{n}{i} = \binom{n}{n-i}$,

(f) $\binom{n}{i} + \binom{n}{i+1} = \binom{n+1}{i+1}$.

3 Construct samples of four observations each such that (a) the sample variance is zero, (b) $\beta_1 = 0$, (c) $\mu_3 < 0$, (d) $\mu_3 > 0$, (e) $\beta_2 < 3$, (f) $\beta_2 > 3$.

# 2
# Introduction to probability

The student of language is often concerned with populations made up of stretches of speech, composed of a small inventory of atomic items (phones, morphs, words, etc.), which vary through time and geographical space. The researcher is interested in making inferences about the nature and the distribution of variates defined on these populations, and in answering such questions as: when are the distributions of two such variates essentially the same? with what degree of certainty can we say their distribution is the same or different? Indeed, language researchers, like other social scientists, are faced with the problem of weighing evidence and making decisions in the face of uncertainty. Often they are faced with a number of conjectures which can be made, given the data, but not enough information to argue deductively to the truth of any one of these conjectures. This position is analogous to that of the gambler who is faced with a number of possible outcomes to the roll of his dice but has not enough information to deduce the final outcome of a particular roll.

In the seventeenth and eighteenth centuries, an interest in gambling and in the new lotteries and insurance companies which were being organized at the time led Blaise Pascal, Jacob Bernoulli, Abraham De Moivre, G.W. Leibniz, and others to propose a theory of probability, from which the modern theory of probability and statistics developed. It was soon discovered that probability theory could be applied in areas of uncertainty other than gambling and lotteries. As we shall see, linguistics and the study of style are two such areas.

Statistics can be characterized, somewhat superficially, as applied probability. This relationship dictates our development here. First we must provide the probabilistic foundations before it is possible to consider the statistics of language.

The term statistics as used in this and succeeding chapters refers to inferential statistics more than to descriptive statistics (see the discussion of the terms in chapter 1).

A cautionary note is in order when applying statistical techniques. It is always wise, before collecting numerical data for analysis, to consult a mathematical statistician. This precaution can save the language researcher great amounts of misspent time and effort.

2.1
PROBABILITY AND RELATIVE FREQUENCY

The theory of probability was developed to study events which contain more or less chance or random components in their makeup. The probability of a particular event is put forward as a numerical measure of the likelihood that the event will occur.

The following simple example will serve as a model for further developments. Consider a (fair) coin which is *flipped*. It will fall to the ground (table or what have you) with either a *head H* or *tail T* up. The flip of the coin can be treated as an *experiment* whose *outcomes* are *H* and *T*.

Suppose we perform this experiment a large number of times, doing, say, $n$ flips

(1)  $F_1, F_2, ..., F_n,$

and note the results. If $f_H(n)$ and $f_T(n)$ denote the number of times a head and tail appear, respectively, in the $n$ flips (1), then the *relative frequency* of heads is

(2)  $f_H(n)/n$

and of tails

(3)  $f_T(n)/n.$

If the coin is fair and if in each experiment the flipping is done properly, then intuitively one feels that as $n \to \infty$ ($n$ increases indefinitely) the ratios (2) and (3) will approach $\frac{1}{2}$. That is

(4)  $\lim_{n \to \infty} [f_H(n)/n] = \lim_{n \to \infty} [f_T(n)/n] = \frac{1}{2}.$

We feel this because the obvious symmetry of the coin ensures that at each toss an $H$ and a $T$ are *equally likely*, and so, over the long run, their respective relative frequencies should be equal. In fact, as you will see if you flip a coin a large number of times, the convergence indicated in (4) is rather slow. This limiting value of the relative frequency of an event is what we shall designate as its *probability*. Thus in the coin flip the probability of $H$ is $\frac{1}{2}$, as is the probability of $T$.

EXAMPLE 1  The convergence of the relative frequencies is sometimes more rapid than for a coin toss. J. Krámský (1966) considered the occurrence of vowels in certain languages whose vowel system contained only the five vowels /a/, /e/, /i/, /o/, /u/. For Czech he took a sample of 1000 words from running text and considered separately the first 40, 80, 120, 200, 240, 320, 500, 700, and 1000 words. His results are given in table 2.1 (table VI in Krámský's paper).

Considering for example /u/, the relative frequencies $f_u(n)/n$ for various

C

**TABLE 2.1**

Vowel frequencies of the first 40, 80, . . . , 1000 words in a 1000-word Czech text

| Vowel | 40 | | 80 | | 120 | | 200 | | 240 | | 320 | | 500 | | 700 | | 1000 | |
|---|---|---|---|---|---|---|---|---|---|---|---|---|---|---|---|---|---|---|
| | Abs. | % | Abs. | % | Abs. | % | Abs. | % | Abs. | % | Abs. | % | Abs. | % | Abs. | % | Abs. | % |
| i | 19 | 24.68 | 35 | 23.18 | 62 | 25.62 | 104 | 26.06 | 127 | 26.29 | 165 | 25.86 | 231 | 24.42 | 321 | 24.14 | 433 | 22.96 |
| u | 5 | 6.49 | 15 | 9.93 | 23 | 9.51 | 33 | 8.27 | 39 | 8.08 | 51 | 8.00 | 77 | 8.14 | 108 | 8.12 | 156 | 8.27 |
| e | 22 | 28.57 | 41 | 27.15 | 63 | 26.03 | 105 | 26.32 | 131 | 27.12 | 176 | 27.59 | 259 | 27.38 | 364 | 27.37 | 506 | 26.83 |
| o | 12 | 15.58 | 26 | 17.22 | 41 | 16.94 | 70 | 17.55 | 82 | 16.98 | 110 | 17.24 | 169 | 17.86 | 253 | 19.02 | 363 | 19.25 |
| a | 19 | 24.68 | 34 | 22.52 | 53 | 21.90 | 87 | 21.80 | 104 | 21.53 | 136 | 21.31 | 210 | 22.20 | 284 | 21.35 | 428 | 22.69 |
| TOTAL | 77 | 100.00 | 151 | 100.00 | 242 | 100.00 | 399 | 100.00 | 483 | 100.00 | 638 | 100.00 | 946 | 100.00 | 1330 | 100.00 | 1886 | 100.00 |

TABLE 2.2

| $n$ | 77 | 151 | 242 | 399 | 483 | 638 | 946 | 1330 | 1886 |
|---|---|---|---|---|---|---|---|---|---|
| $f_u(n)/n$ | 0.0649 | 0.0993 | 0.0951 | 0.0827 | 0.0808 | 0.0800 | 0.0814 | 0.0812 | 0.0827 |

values of $n$ of /u/ in the vowels counted are given in table 2.2. Note that the convergence is quite good for such a small sample of vowels.

In Krámský's study, the underlying experiment involved choosing vowels at random in a text. The outcomes of this experiment are /a/, /e/, /i/, /o/, or /u/. If we assume that (i) Czech texts are homogeneous as regards the frequencies of the various vowels and (ii) the conditioning of one vowel by other nearby vowels is small enough so as not to affect the relative frequencies in large samples, then for example

$$\lim_{n \to \infty} [f_u(n)/n] \approx 0.08$$

seems to constitute a good approximation of the probability that a vowel selected at random from a Czech text would be a /u/.

In his sample of 15,465,010 words of spoken General American, A. Hood Roberts (1965) obtained the relative frequencies for vowels shown in table 2.3.

Assuming an average frequency of even two vowels per word, the relative frequencies in table 2.3 are computed from more than 30 million vowels. If Roberts's corpus is at all representative, then each of the above relative frequencies might be thought to be very near the probability that a vowel selected at random from a General American text would be the associated vowel.

A word about Roberts's methods is perhaps in order. His study is based on Ernest Horn's *A Basic Writing Vocabulary*, which, despite its title, consists of, in Roberts's words, 'something like the "everyday" English of the American adult' (Roberts 1965, p. 27 and p. 28 penultimate paragraph). Using Horn's list, Roberts then had an informant (a speaker of 'General American') put each of these words into a sentence from which the segmental phonemes of the given word were recorded. The relative frequencies of these segmental phonemes were then computed using Horn's relative word-frequencies as a basis.

Let us now consider more closely the idea of relative frequency. We can consider the relative frequency of compound outcomes, for example obtaining a high vowel. The high vowels are /i/ and /u/. The relative frequency of a high vowel in a successive selection of $n$ vowels is $f_i(n) + f_u(n)$ divided by $n$, where $f_i(n)$ is the frequency of i in the $n$ vowel selections and $f_u(n)$ is the frequency of u in

TABLE 2.3

| /ə/ | 0.3273451 | /u/ | 0.0528280 |
|---|---|---|---|
| /i/ | 0.2573291 | /æ/ | 0.0426501 |
| /e/ | 0.1311444 | /o/ | 0.0425611 |
| /a/ | 0.1283451 | /ɔ/ | 0.0177971 |

*the same n* selections. More generally, if $r_A(s_1, s_2, ..., s_n)$ stands for the relative frequency of a vowel from the set $A$ in the $n$ selections $s_1, s_2, ..., s_n$, then for $A = \{i, u\}$ we have

$$r_{\{i,u\}}(s_1, s_2, ..., s_n) = r_{\{i\}}(s_1, ..., s_2) + r_{\{u\}}(s_1, ..., s_n)$$

Indeed,

$$r_{\{i,u\}}(s_1, ..., s_n) = \frac{f_i(n) + f_u(n)}{n} = \frac{f_i(n)}{n} + \frac{f_u(n)}{n}$$

$$= r_{\{i\}}(s_1, ..., s_n) + r_{\{u\}}(s_1, ..., s_n) = 0.257 + 0.053$$

$$= 0.310 \quad \text{(if we round off to three decimal places).}$$

The relative frequency of a front vowel (these are /i/, /e/, and /æ/) in the same sequence of selections is

$$r_{\{i,e,æ\}}(s_1, ..., s_n) = r_{\{i\}}(s_1, ..., s_n) + r_{\{e\}}(s_1, ..., s_n) + r_{\{æ\}}(s_1, ..., s_n)$$

$$= 0.257 + 0.131 + 0.043 = 0.431.$$

The relative frequency of vowels which are both front and high is the relative frequency of obtaining a vowel in the intersection of the sets $\{i, e, æ\}$ of front vowels and $\{i, u\}$ of high vowels:

$$r_{\{i,u\} \cap \{i,e,æ\}}(s_1, ..., s_n) = r_{\{i\}}(s_1, ..., s_n) = 0.257.$$

The relative frequency of a vowel which is either front or high is

$$r_{\{i,u\} \cup \{i,e,æ\}}(s_1, ..., s_n)$$

$$= r_{\{i,u\}}(s_1, ..., s_n) + r_{\{i,e,æ\}}(s_1, ..., s_n) - r_{\{i,u\} \cap \{i,e,æ\}}(s_1, ..., s_n)$$

$$= 0.310 + 0.431 - 0.257 = 0.484.$$

In general, suppose $\mathscr{E}$ is an experiment with a set $S$ of possible outcomes, for example the tossing of a coin or the random selection of an individual from a population. Suppose further that this experiment can be performed an arbitrary number of times to obtain a set $t_1, t_2, ..., t_n$ of trials of $\mathscr{E}$. Let $A$ be some event dependent on the outcome of $\mathscr{E}$, for example obtaining a head in a coin toss or obtaining a high vowel in selecting a vowel at random in the corpus of all General American English. Then the relative frequency $r_A(t_1, t_2, ..., t_n)$, abbreviated $r_A$, of the event $A$ for the trials $t_1, t_2, ..., t_n$ satisfies the following conditions:

(5) $0 \leqq r_A \leqq 1$ for any event $A$,

(6) $r_S = 1$

where $S$ is the event that some one of the possible outcomes occurs,

(7) $r_{A \cup B} = r_A + r_B$ if $A \cap B = \emptyset$,

i.e., if $A$ and $B$ have no outcomes simultaneously favourable to both, then the relative frequency of an outcome occurring which is favourable to either $A$ or $B$ is equal to the relative frequency of $A$ plus the relative frequency of $B$. Two events $A$ and $B$ with no outcomes in common (i.e., $A \cap B = \emptyset$) are said to be *mutually exclusive* events.

Properties (5) and (6) are obviously true, and (7) is easily verified. Indeed, if $A$, $B$ are events for which $A \cap B = \emptyset$, then let $t_{i_1}, t_{i_2}, \ldots, t_{i_k}$ be the trials in which the outcomes are favourable to the event $A$ and let $t_{i_{k+1}}, \ldots, t_{i_m}$ be the trials favourable to $B$. Since $A$ and $B$ are mutually exclusive, the two sets of trials are disjoint. Then the frequencies of the events $A$, $B$, and $A \cup B$ are respectively

$$f_A = k, \quad f_B = m - k, \quad f_{A \cup B} = m.$$

The relative frequency of $A \cup B$ is

$$r_{A \cup B} = \frac{f_{A \cup B}}{n} = \frac{m}{n} = \frac{k + m - k}{n} = \frac{k}{n} + \frac{m - k}{n}.$$

$$= \frac{f_A}{n} + \frac{f_B}{n} = r_A + r_B.$$

Expressions (5), (6), and (7) can be used as a basis for the theory of probability mentioned in the introduction to this chapter.

## EXERCISES

1 Use the results of table 2.3 to find the relative frequency in Roberts's sample of (a) mid vowels, (b) low vowels, (c) central vowels, (d) back vowels, (e) vowels which are either central or low, (f) vowels which are both central and high. For those unfamiliar with the terms 'mid,' 'low', 'high,' 'back,' and 'central,' figure 2.1 shows the various possible positions of the tongue when producing a vowel. For example, in producing æ as in *pass* the tongue is in low front position. For more on tongue positions and vowels see Gleason (1955), chapter 3.

2 Use properties (5), (6), and (7) to prove that

(a) $r_{A'} = 1 - r_A$,

(b) $r_{A \cap B} + r_{A-B} = r_A$,

(c) $r_{A \cup B} = r_A + r_B - r_{A \cap B}$,

for arbitrary events $A$ and $B$. The event $A - B$ can be defined as occurring exactly when the outcome of $\mathscr{E}$ is favourable to $A$ but not to $B$.

3 If an experiment $\mathscr{E}$ consists of the rolling of a fair die with six faces each marked with one of the numbers, 1, 2, ..., 6, what are the possible outcomes of $\mathscr{E}$ and what does your intuition tell you are the probabilities that each of the faces will lie upwards after a given roll of the die?

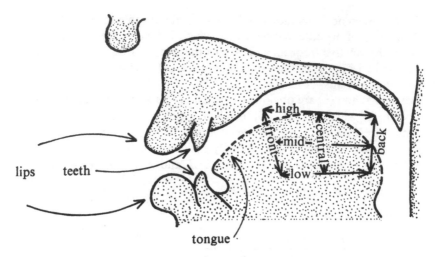

FIGURE 2.1 Positions of the tongue when producing the various English vowels.

## 2.2
PROBABILITY SPACES

Often we would like to make a general statement about a large set of objects when we have access to only a small part of this large set. For example, in chapter 1 we wanted to obtain some idea of the distribution of articles in a work without looking at the entire work. Similarly, in the last section, we hypothesized that A. Hood Roberts's relative frequencies of vowel-types for a 15-million-word sample of General American spoken text were quite close to the relative frequencies (probabilities) for 'all' of General American spoken text. Using the nomenclature introduced in section 1.1, we call the 'set of all' vowels in 'all' spoken General American text the *population* and all the vowels in Roberts's 15 million words a *sample* drawn from that population. I have put quotes around 'all' and 'set of all' in the above sentence because there is a sense in which all spoken text in General American is not well defined. However for the purposes of Roberts's survey and other such studies it is helpful to assume that the population under discussion has some concrete existence.

Another kind of population which might arise in a linguistic discussion is the set $S_0$ of individuals in some community or geographical area. Suppose that among the individuals in $S_0$ the languages $L_1, L_2, ..., L_m$ are spoken. Let $S_i$ be the set of individuals in $S_0$ whose native language (mother tongue) is $L_i$. If we assume that every individual can have only one native language, then the $S_i$ are mutually exclusive, i.e.,

$$S_i \cap S_j = \emptyset$$

when $i \neq j$, and their union is $S_0$, i.e.,

$$\bigcup_{i=1}^{m} S_i = S_0.$$

Greenberg (1956) proposes as a measure of linguistic diversity the probability that a pair of speakers selected at random have the same mother tongue. The population in this case, though more concrete and much closer to the ideal of well-definedness than the vowels in all spoken text in General American, might still be too large to be completely accessible to the researcher. He is then reduced to generalizing from a small (in comparison to the actual population size) sample of the population. As we have noted above, one of the purposes of statistical studies is to make statements about populations based on samples taken from them. Another purpose of statistical theory is to ascertain how much confidence can be placed in such statements.

In order to obtain this statistical theory, we begin by developing an underlying theory of probability. Such a theory can be based on a number of probability models. Any model we choose, and there are many to choose from, should involve two things: (i) a universe of events under discussion, and (ii) a consistent assignment of probabilities to these events.

There are two main schools of thought regarding the philosophical basis upon which a probability theory should be erected – the frequentist and the subjectivist. Adherents of the former believe that probabilities should only be viewed as the *limits of relative frequencies* as they are discussed in section 2.1. The subjectivists, on the other hand, take the view that *probability is an expression of the strength of our knowledge or beliefs*. Except for certain subtle points which need not concern us here, the results obtained from both starting points coincide.

A compelling argument for the subjectivist point of view is that such events as 'Shakespeare wrote Act II, Scene I of *Pericles*' cannot directly be assigned a probability as a limit of relative frequencies.

However, because the frequentist point of view is the easier to explicate and because the most useful statistical tools are based on the frequentist approach, we present our theory of probability and our statistics in a frequentist setting.

The frequentist theory is based on the idea of an experiment $\mathscr{E}$, which can be performed an arbitrary number of times under uniform conditions. Each instance of a performance of this experiment is a *replication* of $\mathscr{E}$. The probability of a certain outcome can be thought of as the limit of the relative frequency of that outcome in a large number of replications.

In most examples, the experiment $\mathscr{E}$ is the selection of an individual from a certain population (cf. section 1.1) and the measurement of the value of some variate $X$ for that individual. For example, $\mathscr{E}$ could be the selection of a 50-word passage at random from the population of all 50-word passages in a certain novel, and the variate $X$ measured could be the number of articles in the passage, as was chosen in section 1.6.

The requirement of performing the experiment $\mathscr{E}$ under uniform conditions can be ensured when sampling from a population by taking a random sample in the sense of section 1.1 each time. This randomness ensures that the sample yields as fair a picture as possible of the population from which it came.

A set $E = \{e\}$ which contains exactly one of the possible outcomes of an experiment $\mathscr{E}$ is often called informally a *simple event* of $\mathscr{E}$. The events mentioned in (i) above can be thought of as sets of outcomes and hence conjunctions of simple events as we shall soon see.

EXAMPLE 1  If $\mathscr{E}$ were the counting of articles in a 50-word passage selected at random from a novel, then the simple events corresponding to this experiment would be the possible values of the variate $X =$ the number of articles in the passage, i.e., the possible numbers of articles: 0, 1, 2, ..., 50.

EXAMPLE 2  If our experiment were tossing a coin and noting the outcome, then there would be only two possibilities: $H$, a head, or $T$, a tail. Hence the simple events in this case are $H$ and $T$.

EXAMPLE 3  If we were selecting at random a real number between 0 and 1 inclusive, the simple events of this experiment would be the set of real numbers between 0 and 1 inclusive, i.e., the set

$$[0, 1] = \{x \in R \,|\, 0 \leqq x \leqq 1\}.$$

EXAMPLE 4  In the mother-tongue example mentioned above, we might perform the experiment of selecting at random an individual in $S_0$ and noting his mother tongue. In this case the simple events would be the possible mother tongues: $L_1, L_2, L_3, ..., L_m$.

In addition to the existence of simple events, we must also, by (ii) above, require that each of these events possess a probability, i.e., a number between 0 and 1 inclusive which reflects its relative frequency of occurrence in a large number of replications of the experiment.

EXAMPLE 1 (continued)  At present we do not have sufficient theoretical framework to establish a probability for each of the simple events in the experiment outlined in example 1. However, we shall come back to this problem later.

EXAMPLE 2 (continued)  Assuming the coin is fair, we note that each of the simple events $H$ and $T$ is equally likely, so the probabilities $P(H)$ and $P(T)$ of $H$ and $T$ respectively are both $\frac{1}{2}$.

EXAMPLE 3 (continued)  In the selection of a point at random from the interval

[0, 1], each point is equally likely. Since there is an infinitude of possible outcomes, $P(\alpha)$, the probability that the point selected is $\alpha$ must be zero.

EXAMPLE 4 (continued)   If $|S_0|$ denotes the number of elements in $S_0$ and the experiment is the selection of an individual at random from $S_0$, then since each individual is as likely to be chosen as any other, the limit of the relative frequencies of the choice of a given individual $s$ in a large number of replications of the experiment should approach

$$P(s) = 1/|S_0|$$

for every $s \in S_0$. If there are $k$ native speakers of $L_1$, then the relative frequencies of $L_1$-native speakers should approach $P(L_1) = k/|S_0|$.

More often than not our interest is not solely engaged with simple events. For example, if we performed the experiment of rolling a die and noting the number on the upper face, we might be interested in the probability of obtaining an outcome which is 'less than 4,' or, as in example 4, we might be interested in obtaining the probability that the randomly selected individual $s \in S_0$ is a native speaker of $L_1$. For these reasons we must develop a theory of compound events. In each case, these compound events should have properties (i) and (ii) mentioned above, and, as well, (iii) they should be conjunctions of simple events.

The satisfaction of properties (i), (ii), and (iii), which is a requirement of both the frequentists and the subjectivists, can be ensured if we require our events and their probabilities to satisfy the system of axioms given below. This system is sufficiently broad to be acceptable by both schools of thought.

First, however, a word about axiom systems in general may be appropriate. Since an axiom system is meant to have application in many areas, it is usual to make it as abstract as possible. In the Euclidean geometry we learned in school, we ensured this by building the axioms in terms of the undefined entities 'point' and 'line' and the undefined relations 'intersection' between lines, 'containment' between a 'point' and a 'line,' and an identity relation between 'points' and 'points' and 'lines' and 'lines.' Thus a typical axiom might be:

'If $P_1$ and $P_2$ are two points and $P_1 \neq P_2$, then there is a unique line $l$ which contains both $P_1$ and $P_2$.'

Any system which satisfies the set of axioms for Euclidean geometry, i.e., which provides a concrete interpretation of the undefined terms 'point' and 'line' and the various relations between them in such a way that the axioms hold true, will also satisfy any deduction made from these axioms.

In an axiom system for probability, the undefined terms are *outcomes, events,* and *probability*. Outcomes and events are required to satisfy the following conditions (axioms):

(1) There is a *universal event* (i.e., an event that always occurs) $U$ which can be characterized as 'any one of the possible outcomes occurs.' Each event is a collection of outcomes – the outcomes *favourable* to it.

(2) The set $E$ of all events includes the universal event $U$, and if $A$ and $B$ are events, then

$A \cup B$, 'either $A$ or $B$ occurs,'

$A \cap B$, 'both $A$ and $B$ occur,'

$A'$,     '$A$ does not occur,'

are all among the events.

As a matter of definition the impossible event $U'$ is designated $\emptyset$. It is, of course, no coincidence that we use set notation for the combination of events. Indeed, our events are sets of outcomes.

Condition (ii) above requires that a probability be defined for each of our events, and if these probabilities are to be limits of relative frequencies, then expressions (5), (6), (7) and exercise 2(a) in section 2.1 indicate that the probability $P(D)$ of an arbitrary event $D$ must be defined, and the following set of axioms satisfied. If $A$, $B$ are arbitrary events, then

(3) $0 \leqq P(A) \leqq 1$;

(4) $P(U) = 1$;

(5) if $A \cap B = \emptyset$, i.e., if $A$ and $B$ are mutually exclusive, then $P(A \cup B) = P(A) + P(B)$.

A set $E$ of events satisfying (1) and (2) together with a probability function $P$ satisfying (3)–(5) constitutes a *probability space*.

As we have already indicated, conditions (1)–(5) are the axioms of a probability space in exactly the same sense as Euclid's axioms are the axioms of geometry. Any object satisfying these conditions is a 'probability space' in our sense. From these conditions, we shall make certain deductions that will then hold in any probability space. The virtue of this framework is that the deductions need be made only once for the abstract 'probability space'; they then become true *a fortiori* for any specific concrete probability space in virtue of the fact that it satisfies the axioms. Note that the collection $E$ of events postulated by the axioms (1) and (2) is not required to include the simple events defined earlier. It is sometimes useful to allow for probability spaces where the one-outcome sets are not among the events under consideration. However, in nearly all cases we consider here the collection $E$ includes all simple events. Exercise 6 at the end of this section represents a case where the simple events are not included in $E$.

The probability space with axioms (1)–(5) is, in some cases, too general for certain purposes and hence we sometimes require a more restrictive concept. A probability space is *continuous* if

(6)  when $A_1$, $A_2$, $A_3$, ... belong to $E$, then $E$ also contains

$\bigcup A_n$, 'one of $A_1$, $A_2$, ... occurs,' and

$\bigcap A_n$, 'each of $A_1$, $A_2$, ... occurs,' and

(7)  when $A_1$, $A_2$, $A_3$, ... belong to $E$, and $A_i \cap A_j = \emptyset$ for all $i \neq j$, then[1]

$$P(\bigcup A_n) \overset{d}{=} P(A_1 \cup A_2 \cup ...) = \sum_{n=1}^{\infty} P(A_n) \overset{d}{=} P(A_1) + P(A_2) + ... .$$

The need for continuity is purely technical and is of little interest to us. I have included its definition for the sake of completeness only. Having done so, I must also add some further remarks.

(i) If the set of outcomes is a finite set and the set $E$ of events includes the simple events, then it can be shown that the set of events must be the set of all subsets of $S$ and that (6) and (7) are automatically true. So if a probability space has only a finite set of simple events, it is automatically a continuous probability space.

(ii) In the more advanced literature the term probability space generally refers to what we have called a continuous probability space. However, for our present considerations we can get along with the simpler concept of probability space in our more general sense, i.e., a set of events and a probability satisfying (1), (2), (3), (4), and (5).

EXAMPLE 2 (continued)  For the coin-tossing experiment the set of possible outcomes is $S_c = \{H, T\}$. If (1) and (2) are to hold, then the set $E_c$ of events must contain

$$U = \{H\} \cup \{T\}, \quad \{H\} \cap \{T\} = \emptyset, \quad \{H\}' = \{T\}, \quad \{T\}' = \{H\}.$$

Therefore

$$E_c = \{\emptyset, \{H\}, \{T\}, U\}.$$

Since $H$ and $T$ are equally likely, we have

$$P_c(\{H\}) = P_c(\{T\}) = \tfrac{1}{2}.$$

By (4), we must have $P_c(U) = 1$. To obtain a value for $P(\emptyset)$, note that

$$\{H\} \cup \emptyset = \{H\} \quad \text{and} \quad \{H\} \cap \emptyset = \emptyset.$$

Therefore, by (5),

$$\tfrac{1}{2} = P_c(\{H\}) = P_c(\{H\} \cup \emptyset) = P_c(\{H\}) + P_c(\emptyset) = \tfrac{1}{2} + P_c(\emptyset).$$

and so $P_c(\emptyset) = 0$. We shall see shortly that $P_c(\emptyset) = 0$ is deducible directly from the axioms for a probability space. We have just established that we must have

---

1  The symbol $\overset{d}{=}$ means 'is defined as equal to.'

(8)  $P_c(\emptyset) = 0, \quad P_c(\{H\}) = P_c(\{T\}) = \frac{1}{2}, \quad P_c(U) = 1.$

This assignment of probabilities is intuitively acceptable and can easily be seen to satisfy (4), (5), and (6). Thus $E_c$ and $P_c$ as defined above constitute a probability space.

EXAMPLE 1 (continued)   In this case the collection of outcomes is

$$S_A = \{1, 2, 3, ..., 50\},$$

the possible numbers of articles in a 50-word passage. Then the events must include $\emptyset$; the simple events $\{1\}, \{2\}, ..., \{50\}$; the two-element events (conjunctions of simple events) $\{1, 2\}, \{1, 3\}, ..., \{49, 50\}$; the three-element events (conjunctions of simple events and two-element events) $\{1\} \cup \{2, 3\} = \{1, 2, 3\}$, $\{1, 3, 4\}, ..., \{48, 49, 50\}$; and so on up to the universal events $\{1, 2, 3, ..., 50\}$ $= U$. Thus the set $E_A$ of events in this case contains all possible subsets of the collection of outcomes; there will be $2^{50}$ of these.

Now suppose that it were possible to assign a probability to each of the possible numbers of articles in a 50-word passage, i.e., suppose the probability of obtaining $k$ articles is $p_k$. Then $P_A(\{k\}) = p_k$ for $k = 1, 2, ..., 50$ and we would set

$$P_A(\{j, k\}) = p_j + p_k$$

if $j \neq k$ in order to satisfy (5). It is not difficult to show that (5) requires that if $B = \{i_1, i_2, ..., i_m\}$ is an arbitrary subset of $S_A$ ($i_j \neq i_k$ when $j \neq k$), then

(9)  $P_A(B) = \sum_{k=1}^{m} p_{i_k}.$

It should be clear to the reader that if (9) holds, $E_A$ and $P_A$ constitute a probability space.

EXAMPLE 4 (continued)   Now we have two alternatives depending on the problems we wish to discuss. First, if our only interest is in the native language of a given speaker, we might choose as our set of outcomes $S_N = \{L_1, L_2, ..., L_m\}$ and proceed as follows. If $S_1, S_2, ..., S_m$ are the subsets of $S_0$ composed of exactly the native speakers of languages $L_1, L_2, ..., L_m$ respectively, then, as we have already observed, $S_i \cap S_j = \emptyset$ for $i \neq j$, because an individual is assumed to have only one native language, and $\bigcup_{k=1}^{m} S_k = S_0$, because everyone in $S_0$ has a native language. We have also observed that it is natural to assign the probability $p_i = |S_i|/|S_0|$ to the simple event that a speaker selected at random will be a native speaker of $L_i$. Remember $|A|$ stands for the number of elements in the set $A$. As in examples 1 and 2, the set $E_N$ of events turns out to be the set of all subsets of $S_N$, and for a particular subset $B$ of native languages $P_N(B)$ is the sum of the $p_i$ corresponding to the $L_i$ in $B$. The definitions of $E_N$ and $P_N$ clearly fit

(1)–(5) and are intuitively satisfying as well. Thus $E_N$ and $P_N$ constitute a probability space.

As for the second alternative, suppose the individuals of $S_0$ have other properties of interest besides their native language. For example, some may be speakers of one or more languages in addition to their native language. It is then advantageous to take the selection of each individual to be a simple event so that the set of simple events is $S_0$; then if $B = \{s_1, \ldots, s_q\}$ is a set of individuals, the probability of selecting at random an individual in $B$ would intuitively seem to be

(10) $P_0(B) = |B|/|S_0|$.

The reader should be able to justify for himself that if $E_0$ is taken to be the set of subsets of $S_0$, then $E_0$ and $P_0$ constitute a probability space.

EXAMPLE 3 (continued)  Selecting a point in [0, 1] at random presents certain technical problems because the set of simple events is an infinite set where every such event has a natural probability zero. The sort of events we would like to consider include not only points in [0, 1] but also intervals such as

$$[\alpha, \beta] = \{x \in [0, 1] \mid \alpha \leq x \leq \beta\}.$$

There is no immediate way of assigning a probability to the selection of a point in the interval $[\alpha, \beta]$ using the knowledge that the probability of each $x \in [\alpha, \beta]$ (i.e., the probability of selecting such an $x$) is zero. However, when selecting points at random from [0, 1] it is quite natural to assign to the event of a selection of a point from $[\alpha, \beta]$ the probability

$$P([\alpha, \beta]) = \frac{\beta - \alpha}{1 - 0} = \beta - \alpha,$$

which is the ratio of the length of $[\alpha, \beta]$ to the length of the total interval [0, 1]. In more advanced texts it is possible to show that a unique smallest collection $E_{[0, 1]}$ of subsets of [0, 1] exists which satisfies (1), (2), *and* (6) and that a probability $P$ can be found which satisfies (3), (4), (5), *and* (7) and which assigns to every $x$ in [0, 1] a probability zero and to every interval $[\alpha, \beta]$ the probability

$$P([\alpha, \beta]) = \beta - \alpha.$$

Thus the pair $E_{[0, 1]}$ and $P$ constitute a continuous probability space. It can be shown that $E_{[0, 1]}$ is not the set of all subsets of [0, 1] and that no $P$ exists with properties (3), (4), (5), and (7) if the set of compound events is taken to be the set of all subsets of [0, 1].

Now let us proceed to make certain deductions from the axioms for a probability space. The results obtained will of course hold for any arbitrary probability space.

THEOREM 1    Let $E$ and $P$ constitute a probability space. Then

(11) $P(\emptyset) = 0.$

If $A$ and $B$ belong to $E$, then

(12) $P(A') = 1 - P(A)$

and

(13) $P(A \cup B) = P(A) + P(B) - P(A \cap B).$

The proof of (11) follows the lines indicated by the remarks in the continuation of example 2. First note than $A \cap \emptyset = \emptyset$ and $A \cup \emptyset = A$ for any event $A$. Then by axiom (5)

$$P(A \cup \emptyset) = P(A) + P(\emptyset),$$

but $P(A) = P(A \cup \emptyset)$. Thus

$$P(A) = P(A) + P(\emptyset).$$

Subtracting $P(A)$ from both sides yields (11). Expression (12) has a proof which is almost identical in form with that of (11). First note that $A \cap A' = \emptyset$ and $A \cup A' = U$; then (5) yields $P(A) + P(A') = P(U) = 1$ and so (12) follows immediately by subtracting $P(A)$ from both sides.

The proof of (13) can be constructed easily from the relations illustrated in the Venn diagram given in figure 2.2.

EXAMPLE 5    Another example will illustrate some of the results of theorem 1. Consider $E_0$ and $P_0$ again. Allowing for bilingual speakers, let $A_1$ and $A_2$ be the set of speakers of $L_1$ and $L_2$ respectively. Since there may be bilinguals, the set $A_1 \cap A_2$ need not be void. If we wish to find the probability of selecting a speaker of either $L_1$ or $L_2$ from $S_0$, we need the value of $P(A_1 \cup A_2)$, which, by theorem 1, is

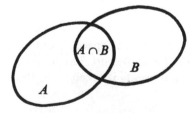

FIGURE 2.2   $A \cup B = A \cup (B - A \cap B)$
where $A \cap (B - A \cap B) = \emptyset$, and
$(B - A \cap B) \cup (A \cap B) = B$ where
$(B - A \cap B) \cap (A \cap B) = \emptyset.$

$$P(A_1 \cup A_2) = P(A_1) + P(A_2) - P(A_1 \cap A_2)$$
$$= \frac{|A_1| + |A_2| - |A_1 \cap A_2|}{|S_0|}.$$

The probability of obtaining a non-speaker of $L_1$ is

$$P(A_1') = 1 - P(A_1) = 1 - \frac{|A_1|}{|S_0|} = \frac{|S_0| - |A_1|}{|S_0|},$$

also by theorem 1.

## EXERCISES

1 Let the set $S$ of outcomes of an experiment be finite, let $E$ be the set of all subsets of $S$, and for $A \in E$ let $P(A) = |A|/|S|$. Show that axioms (1)–(5) hold when $E$ is taken to be the events and $P$ the probability function.

2 Let $S$ be a finite set, $E$ be the set of all subsets of $S$, and suppose that for each $x \in S$

(*) $P(\{x\}) = p_x$.

For $A = \{a_1, a_2, ..., a_k\} \in E$, let

(**) $P(A) = \sum_{i=1}^{k} p_{a_i}$.

Show that the probability function $P$ defined on the set $E$ of events by (**) is the only one which satisfies axioms (3), (4), and (5) *and also* (*).

3 If $A$ and $B$ are events, show that $A \subseteq B$ implies $P(A) \leq P(B)$, where $A \subseteq B$ holds exactly when the occurrence of $A$ implies the occurrence of $B$, i.e., when every outcome favourable to $A$ is also favourable to $B$. (Hint: show that $A = A \cap B$ when $A \subseteq B$.)

4 If $A$, $B$, and $C$ belong to $E$, show that

$$P(A \cup B \cup C) = P(A) + P(B) + P(C) - P(A \cap B) - P(A \cap C) - P(B \cap C)$$
$$+ P(A \cap B \cap C).$$

5 Show that for events $A_1, A_2, A_3$, if $A_1 \cap A_2 = \emptyset$, $A_1 \cap A_3 = \emptyset$, and $A_2 \cap A_3 = \emptyset$, then $P(A_1 \cup A_2 \cup A_3) = P(A_1) + P(A_2) + P(A_3)$.

6 A natural probability space to associate with vowels in General American speech is the following. Let the set of outcomes $S$ be the set of all vowel instances in General American speech, let $E$ contain the sets

$$A_\text{ə} = \{x \in S \mid x \text{ is an e}\}, \quad A_\text{i} = \{x \in S \mid x \text{ is an i}\},$$

and analogously for $A_\text{e}$, $A_\text{a}$, $A_\text{u}$, $A_\text{æ}$, $A_\text{o}$, $A_\text{ɔ}$, and let $P(A_\text{ə}) = 0.3273451$, etc., using table 2.3. Since the collection $A_\text{ə}$, $A_\text{i}$, ..., $A_\text{ɔ}$ is a mutually disjoint set of sets whose union is $S$, each of the $2^8 = 256$ unions (including the 'empty' union) can be assigned a probability compatible with the axioms) in a natural way by letting, for example,

$$P(A_\vartheta \cup A_i \cup A_u) = P(A_\vartheta) + P(A_i) + P(A_u).$$

Thus if we let $E$ be the set of all unions which can be formed from the eight sets $A_\vartheta$, $A_i$, ..., $A_\vartheta$, then every set in $E$ has a probability. (a) Do $E$ and $P$, in this case, constitute a probability space? (b) Verify that $P(S) = 1$ by finding $P(A_\vartheta) + P(A_i) + ... + P(A_\vartheta)$. (c) What is $P(A_\vartheta')$?

## 2.3

### INDEPENDENCE AND CONDITIONAL PROBABILITY

In section 2.1 we discussed A. Hood Roberts's sample frequencies-of-occurrence of each of the vowels of General American speech relative to the frequency of all such vowels in his sample. Roberts also gives relative frequencies for all the phonemes of General American. In the former, the population under consideration is composed of the vowels occurring in 'all' spoken General American, and, in the latter, the population is composed of the phonemes in 'all' spoken General American.

Suppose $G$ stands for the population composed of all particular instances (tokens if you like) of phonemes occurring in 'all' spoken General American. These will be instances of /ə/, /i/, /t/, /y/, ..., /ž/ (the types); Roberts lists 32. We can let the experiment be the selection at random of one of these possible types, so that the set of outcomes is

$$S_G = \{/ə/, /i/, /t/, /y/, ..., /ž/\}.$$

Let the set of events $E_G$ be the set of all subsets of $S_G$; there will be $2^{32}$ such sets. Roberts (1965) gives the relative frequencies of each of the phonemes (appendix II), which we shall take to be probabilities. From his vowel entries we have drawn up table 2.4.

If we identify the relative frequencies that Roberts obtained for these events with their probabilities, then we have $p_\vartheta = 0.118$, $p_i = 0.093$, ..., $p_\vartheta = 0.006$ after rounding off to three decimal places. If for an arbitrary element $A$ of $E_G$ we assign the probability

$$P_G(A) = \text{sum of all } p_k \text{ such that } k \in A$$

where $k$ takes the values /ə/, /i/, /t/, /y/, ..., /ž/, then $E_G$ and $P_G$ constitute a prob-

TABLE 2.4

| | | | |
|---|---|---|---|
| /ə/ | 0.1181959 | /u/ | 0.0190748 |
| /i/ | 0.1929149 | /æ/ | 0.0153998 |
| /e/ | 0.0473529 | /o/ | 0.0153677 |
| /a/ | 0.0463421 | /ɔ/ | 0.0064261 |

ability space. Note, however, that the values in table 2.3 differ from those in table 2.4. This difference is explained as follows. By adding the entries of table 2.4, we obtain the result that the relative frequency of a vowel in Roberts's sample from $G$ is 0.361, when rounded off to three places. We also find that the ratio of each entry in table 2.4 to the corresponding entry in table 2.3 is about[2] 0.361. If $n$ and $m$ are, respectively, the number of occurrences of phonemes and the number of occurrences of vowels in Roberts's sample, then, for example, if $f_{\vartheta}(n)$ is the number of occurrences of $/\vartheta/$, the relative frequency of $/\vartheta/$ among all the phonemes is

(1) $\dfrac{f_{\vartheta}(n)}{n} = \dfrac{m}{n}\dfrac{f_{\vartheta}(n)}{m}$

where $f_{\vartheta}(n)/n$ appears in table 2.4 and $f_{\vartheta}(n)/m$ appears in table 2.3. The quotient $m/n$ is the relative frequency of a vowel. Equation (1) can be interpreted as follows. The relative frequency $r_{\vartheta}$ of $/\vartheta/$ in Roberts's sample is equal to the relative frequency $r_V$ of a vowel times the 'conditional' relative frequency $r_{\vartheta|V}$ of $/\vartheta/$ among the vowels. Thus, knowing the relative frequency $r_{\vartheta|V}$ of $/\vartheta/$ among the vowels in Roberts's sample and the relative frequency of vowels $r_V$, we obtain the relative frequency $r_{\vartheta}$ of $/\vartheta/$ among all phonemes:

(2) $r_{\vartheta} = r_V r_{\vartheta|V}.$

Now let us return to the probability space of $E_G$ and $P_G$. Since our intuitive notion of probability is the limit of relative frequencies, we might rewrite (2) in terms of probabilities as

(3) $P(\{/\vartheta/\}) = P(V)P(\{/\vartheta/\}|V),$

which is to say that the probability of selecting (at random) an instance of $/\vartheta/$ is equal to the probability of selecting a vowel (i.e., an element in $V = \{/\vartheta/\} \cup \{/i/\} \cup \ldots \cup \{/\mathfrak{o}/\}$) times the *conditional probability of selecting $/\vartheta/$ knowing that it is a vowel that is being selected.*

This notion of conditional probability can easily be generalized to arbitrary probability spaces and will prove useful in the sequel. Let $E$ and $P$ constitute an arbitrary probability space. Then for any pair of events $A$, $B$ in $E$ such that $P(B) \neq 0$, the *conditional probability of $A$ given $B$* is defined to be

(4) $P(A|B) = P(A \cap B)/P(B).$

First, it might be well to note that a conditional probability is indeed a probability.

---

2 Actually if the entries of table 2.4 are divided by the corresponding entries in table 2.3 the ratio is not exactly 0.361. This is due to error incurred through the rounding of the decimal expressions involved in the calculations, both by us and by Roberts. The differences between the ratios and 0.361 will be seen to be greater for the less frequent vowel. This is because relatively more information is lost through round-off in these cases.

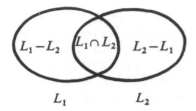

$$L_1 \qquad\qquad L_2$$

FIGURE 2.3   The region on the left represents the set of speakers of $L_1$ and that on the right the set of speakers of $L_2$. The remaining subregions are named appropriately.

THEOREM 1   If $E$ and $P$ satisfy axioms (1)–(5) in section 2.2 and hence constitute a probability space and if $P(B) \neq 0$, then $E$ and $P(\cdot|B)$ also satisfy axioms (1)–(5).

REMARK   We use $P(\cdot|B)$ to stand for the function defined on $E$ whose values are $P(A|B)$ for each $A \in E$.

PROOF   Since $E$ already satisfies expressions (1) and (2) in section 2.2, we need only consider (3)–(5). We verify axioms (3) and (4) of section 2.2 and leave (5) to the reader. Since for arbitrary $A \in E$

$$0 \leq P(A \cap B) \leq P(B)$$

follows from exercise 3 of section 2.2, and since division by $P(B) > 0$ preserves the inequalities, we have

$$0 \leq \frac{P(A \cap B)}{P(B)} \leq \frac{P(B)}{P(B)} = 1,$$

and so axiom (3) holds. If $A = U$, $P(A \cap B) = P(B)$, so $P(U|B) = 1$ and axiom (2) holds.

EXAMPLE 1   Suppose that we have a hypothetical community in which two languages $L_1$, $L_2$ are spoken and that we know the probability that an individual can speak $L_1$ given that he can speak $L_2$. Let it be

$$P(L_1|L_2) = a.$$

Suppose also that we know the probability $P(L_1) = a_1$ that an individual can speak $L_1$, and the probability $P(L_2) = a_2$ that he can speak $L_2$. Suppose we have no further information, but would like to obtain (i) the probability $P(L_1 \cap L_2)$ that an individual can speak both $L_1$ and $L_2$, (ii) the (conditional) probability $P(L_2|L_1)$ that a speaker of $L_1$ can also speak $L_2$, and (iii) the probabilities $P(L_1 - L_2)$ and $P(L_2 - L_1)$ that an individual is a monolingual speaker of $L_1$ and $L_2$ respectively (see figure 2.3).

Using the probability space of examples 4 and 5 in section 2.2, and using the symbols $L_1$ and $L_2$ for the set of speakers of $L_1$ and $L_2$ respectively, we can compute

$$P(L_1 \cap L_2) = P(L_2)P(L_1|L_2) = aa_2$$

from our data (see figure 2.3). We also know that

$$P(L_1 \cap L_2) = P(L_1)P(L_2|L_1)$$

by the definition of conditional probability. Therefore

$$aa_2 = a_1 P(L_2|L_1)$$

and so

$$(5) \quad P(L_2|L_1) = \frac{P(L_2)P(L_1|L_2)}{P(L_1)} = \frac{a_2 a}{a_1}.$$

To find the probabilities of monolinguals, note that

$$L_1 = (L_1 \cap L_2) \cup (L_1 - L_2)$$

and that

$$(L_1 \cap L_2) \cap (L_1 - L_2) = \emptyset.$$

Therefore

$$P(L_1) = P(L_1 \cap L_2) + P(L_1 - L_2)$$

and so

$$P(L_1 - L_2) = P(L_1) - P(L_1 \cap L_2) = a_1 - aa_2.$$

A similar argument yields

$$P(L_2 - L_1) = P(L_2) - P(L_1 \cap L_2) = a_2 - aa_1.$$

Expression (5) is a special case of *Bayes's Rule*: *If $P(A) > 0$ and $P(B) > 0$, then*

$$(6) \quad P(A|B) = P(A)P(B|A)/P(B).$$

This form of Bayes's Rule can easily be generalized as follows: *If $P(A_1 \cup A_2 \cup \ldots \cup A_n) = 1$ and $A_i \cap A_j = \emptyset$ when $i \neq j$ and if $P(B) > 0$ and $P(A_i) > 0$ for each i, then*

$$(7) \quad P(A_j|B) = P(B|A_j)P(A_j) \bigg/ \sum_{i=1}^{n} P(B|A_i)P(A_i).$$

Expression (7) reduces to (6) when $n = 2$ and $A_1 = A$ and $A_2 = A'$, whence (7) becomes

$$P(A_1|B) = \frac{P(A_1)P(B|A_1)}{P(B|A_1)P(A_1)+P(B|A_2)P(A_2)}$$

$$= \frac{P(A)P(B|A)}{P(B|A)P(A)+P(B|A')P(A')} = \frac{P(A)P(B|A)}{P(B)},$$

because

(8)  $P(B|A)P(A)+P(B|A')P(A') = P(B \cap A)+P(B \cap A') = P(B)$

by the definition of conditional probability and axiom (5) in section 2.2.

Intuitively, we feel that two events are independent if the knowledge that one of the events has occurred neither increases nor decreases the likelihood of the second event's occurrence. For example, in tossing a fair coin two successive times, the outcome ($H$ or $T$) on the second toss, we feel, is independent of the outcome on the first toss. On the other hand, the eye-colour of a child is not independent of the eye-colours of its parents. This intuitive discussion is formally embodied in the following definitions. Two events $A$ and $B$ are *independent* if $P(A \cap B) = P(A)P(B)$. Indeed if $P(A) \neq 0$ and $P(B) \neq 0$, then

$$P(A|B)P(B) = P(A \cap B) \quad \text{and} \quad P(B|A)P(B) = P(A \cap B)$$

by the definition of conditional probability. If $A$ and $B$ are independent, the right sides of both these equations can be written $P(A)P(B)$ and if we cancel like non-zero factors, the equations yield

$$P(A|B) = P(A) \quad \text{and} \quad P(B|A) = P(B),$$

so, roughly speaking, if two events are independent, then knowing that one of the events has occurred has no effect on the probability of the occurrence of the other.

In a similar fashion we can show that if two events $A$ and $B$ with non-zero probability are independent, then (and only then) $P(A|B') = P(A)$ and $P(B|A') = P(B)$. Thus the non-occurrence of one of a pair of independent events also does not affect the likelihood of occurrence of the other. We shall meet this concept of independence at a number of crucial points in the succeeding chapters.

EXAMPLE 2  In Markov (1913), the author classified (approximately) 20,000 successive letters in Pushkin's *Eugene Onegin* as $v$ (vowel) or $c$ (consonant) and tabulated the frequencies of overlapping sequences of length three. His data, which consisted of the eight frequencies given in table 2.5, might be assumed to give estimates of the probabilities of the various triples of vowels and consonants in Russian epic poetical writing of the time, if not Russian literature in general. To associate a probability space with these data, consider the population composed of the set of all triples $xyz$ (including overlappings) of successive letters in, say, the corpus of all Russian epic poetry of Pushkin's time. The simple events of

TABLE 2.5

| | | | |
|---|---|---|---|
| $f_{vvv}$ = | 115 | $f_{cvv}$ = | 989 |
| $f_{vvc}$ = | 989 | $f_{cvc}$ = | 6545 |
| $f_{vcc}$ = | 4212 | $f_{ccv}$ = | 3322 |
| $f_{vcv}$ = | 3322 | $f_{ccc}$ = | 505 |

interest to Markov were the various possible combinations of vowels and consonants when a triple of successive letters was selected at random from the poem. Thus in this case

$$S = \{vvv, vvc, vcv, cvv, ..., ccc\}.$$

In order to ensure that $S$, $E$, and $P$ constitute a probability space, we take $E$ to be the set of all subsets of $S$ ($2^8 = 256$ in all). Markov's data can be used to estimate the probabilities $p_{vvv}, p_{vvc}, ..., p_{ccc}$ of the simple events $vvv, vvc, ..., ccc$. Then for $A \in E$ define $P(A)$ as the sum of probabilities corresponding to the simple events contained in $A$.

For example the probabilities of $vvv$ and $vvc$ are estimated to be

$$p_{vvv} = \frac{115}{20,000} = 0.006$$

and

$$p_{vvc} = \frac{989}{20,000} = 0.049.$$

The sample includes overlapping triples, i.e., if $... x_1x_2x_3x_4x_5 ...$ is a processed stretch in the corpus (so that $x_i$ is either $c$ or $v$), then the triples $x_1x_2x_3, x_2x_3x_4, x_3x_4x_5$ among others belong to $U$. Thus the probability of two successive vowels, for example, can be computed from two different events

$$A_{.vv} = \{vvv\} \cup \{cvv\},$$

$$A_{vv.} = \{vvv\} \cup \{vvc\}.$$

In either case, we obtain

$$P(A_{.vv}) = p_{cvv} + p_{vvv} = 0.055,$$

$$(9) \quad P(A_{vv.}) = p_{vvc} + p_{vvv} = 0.055.$$

The probability of the occurrence of a vowel can be computed from any of the following three events:

$$A_{v..} = \{vvv\} \cup \{vvc\} \cup \{vcv\} \cup \{vcc\},$$

$$A_{.v.} = \{vvv\} \cup \{cvv\} \cup \{vvc\} \cup \{cvc\},$$

$$A_{..v} = \{vvv\} \cup \{cvv\} \cup \{vcv\} \cup \{ccv\},$$

all of which yield

$$P(V) = P(A_{v..}) = P(A_{.v.}) = P(A_{..v}) = 0.432$$

and

$$P(V') = P(C) = P(A_{c..}) = P(A'_{v..}) = 1-0.432 = 0.568.$$

To obtain the conditional probability of a vowel following two vowels, form

$$P(A_{..v} \mid A_{vv.}) = \frac{P(A_{..v} \cap A_{vv.})}{P(A_{vv.})} = \frac{P(\{vvv\})}{P(A_{vv.})}$$

$$= \frac{f_{vvv}/20{,}000}{(f_{vvc}+f_{vvv})/20{,}000} = 0.104.$$

Suppose we are asked to compute the probability of a vowel given that a consonant has occurred. This can be computed, for example, as follows:

$$P(A_{.v.} \mid A_{c..}) = \frac{P(A_{cv.})}{P(A_{c..})} = 0.663.$$

Similarly, the probability of a vowel following a vowel is

$$(10) \quad P(A_{.v.} \mid A_{v..}) = \frac{P(A_{vv.})}{P(A_{v..})} = 0.128.$$

Thus it is clear that successive letters are *not* independent, for if they were, then, for example, we would have $P(A_{vv.})$ equal to

$$P(A_{v..}) \cdot P(A_{.v.}) = [P(V)]^2 = 0.187.$$

However, in fact, from (9) we have

$$P(A_{vv.}) = 0.055.$$

For more on transition probabilities connected with linguistic studies see Hultzen et al. (1964).

On occasion we shall need to discuss the independence of more than two events. Suppose $\mathscr{C}$ is a set of events. We say that $\mathscr{C}$ is an *independent* set of events if for every natural number $k \geq 2$, every subset $\{A_1, A_2, ..., A_k\}$ of $k$ different events in $\mathscr{C}$ satisfies the equation

$$P(A_1 \cap A_2 \cap ... \cap A_k) = P(A_1) \cdot P(A_2) \cdot ... \cdot P(A_k).$$

Thus the set $\{A_1, A_2, A_3\}$ of three events is independent provided

$$P(A_1 \cap A_2 \cap A_3) = P(A_1) \cdot P(A_2) \cdot P(A_3),$$

$$P(A_1 \cap A_2) = P(A_1) \cdot P(A_2),$$

$$P(A_1 \cap A_3) = P(A_1) \cdot P(A_3),$$

and

$$P(A_2 \cap A_3) = P(A_2) \cdot P(A_3)$$

all hold.

## EXERCISES

1 If $A$, $B$, $C$ are independent events, show that $A \cup B$ and $C$ are independent.

2 Show that if $P(C) \neq 0$, $P(A \cup B|C) = P(A|C) + P(B|C) - P(A \cap B|C)$.

3 If $A$ and $B$ are independent events, then which of the following pairs of events are independent? (a) $A$, $B'$; (b) $A'$, $B'$; (c) $A$, $A \cap B$; (d) $A'$, $B$; (e) $A$, $S$; (f) $A$, $\emptyset$.

4 Compute the following probabilities using the data of example 2: (a) $P(A_{..v}|A_{vv.})$, (b) $P(A_{..v}|A_{cv.})$, (c) $P(A_{..v}|A_{cc.})$, (d) $P(A_{..v}|A_{vc.})$, (e) $P(A_{..v}|A_{.c.})$, (f) $P(A_{..v}|A_{.v.})$. Is it reasonable to assume that the probability of a vowel depends only on the value ($c$ or $v$) of the letter which immediately precedes it?

5 Consider the hypothetical population of all English word-tokens and the hypothetical experiment of drawing a word at random from this population. Let $A$ stand for the simple event of drawing an English word of known Anglo-Saxon origin and $B$ the event of drawing an instance of a word which is among the tenth of the words sampled in Roberts (1965) which occur most frequently. From Roberts's data we obtain

$$P(A) = 0.7759 \quad \text{(p. 79)},$$

$$P(B) = \frac{1.393 \times 10^6}{1.5465 \times 10^7} \quad \text{(pp. 111--12)},$$

$$P(A|B) = 0.8276 \quad \text{(p. 69)}.$$

Find $P(B|A)$.

6 Let $S_i$ be the event of drawing an $i$-syllable word at random from General American English and let $Q_j$ be the event of drawing at random a word which contains $j$ phonemes from General American English. In Roberts (1965), the sample of 15,465,010 words contains

2,637,562 words belonging to $S_2$,

1,007,061 words belonging to $Q_6$,

and

915,837 words belonging to $Q_6 \cap S_2$.

Find (a) $P(Q_6)$, (b) $P(S_2)$, (c) $P(Q_6 \cap S_2)$, (d) $P(Q_6|S_2)$, (e) $P(S_2|Q_6)$, (f) $P(S_2|Q_6')$, (g) $P(Q_6|S_2')$. Are the events $Q_6$ and $S_2$ independent?

TABLE 2.6

| Syllable type | Relative frequency |
|---|---|
| cvc | 0.335 |
| cv | 0.218 |
| vc | 0.203 |
| v | 0.097 |
| cvcc | 0.078 |
| ccvc | 0.028 |
| vcc | 0.028 |
| ccv | 0.008 |
| ccvcc | 0.005 |

7 Two coins are tossed. Find the probability of obtaining (a) two heads, (b) two tails, (c) a head and a tail.

8 If a coin is tossed three times and the upper face noted each time, construct a probability space corresponding to this experiment and find the probability that (a) $E_1 \equiv$ a head on the first toss, (b) $E_2 \equiv$ a head on the second toss, (c) $E_3 \equiv$ a head on the third toss, (d) there is one head in the three tosses, (e) there are two heads in the three tosses, (f) there are three heads in the three tosses. Are the three events $E_1$, $E_2$, and $E_3$ independent?

9 Prove that if $A$, $B$, $C$ are three events such that $P(A \cap B) \neq 0$, then

$$P(A \cap B \cap C) = P(A)P(B|A)P(C|A \cap B).$$

10 In a study of the telephone speech of u.s. businessmen by Miller (1951), p. 88, it was shown that the syllable-types in table 2.6 were used with the relative frequencies indicated. If we interpret those relative frequencies as probabilities, then: (a) What is the probability of a vowel-initial syllable and what is the probability of a consonant-initial syllable? (b) What is the probability of obtaining a syllable-type cvc given that we have chosen a consonant-initial syllable? (c) What is the probability that vowels and consonants alternate in the syllable?

11 If $A \cup B \cup C = U$ and $A$, $B$, and $C$ are mutually exclusive events, then show that

$$P(D) = P(D|A)P(A) + P(D|B)P(B) + P(D|C)P(C).$$

This is a direct generalization of (8) above.

## 2.4
### URN MODELS

Most readers will have had experience with lotteries, where identical capsules containing the names of the contestants are placed in a large jar or *urn* and after vigorous mixing the winning capsule is drawn. From our previous discussions it should be clear that if the capsules are identical and the mixing was properly done, then owing to the symmetry of the situation the probability of any one of the contestants' (=capsules) winning should be one divided by the total number of contestants.

This analogy of identical capsules (or balls) being drawn from an urn makes a useful intermediate model for the application of probability to various practical situations.

EXAMPLE 1 The use of an intermediate urn model can be trivially illustrated in connection with the calculation of Greenberg's measure of diversity of a linguistic community mentioned in section 2.2. This measure consists of the probability that two individuals selected at random from the community $S_0$ have the same mother tongue. Suppose there are $m$ languages involved, $L_1, L_2, ..., L_m$, spoken (natively) by $n_1, n_2, ..., n_m$ people, respectively. Then the following model can be used to represent the population. Consider a large urn containing

$$n = \sum_{i=1}^{m} n_i$$

balls (all identical to the touch), $n_i$ being of colour $c_i$ for $i = 1, 2, ..., m$ where of course $c_i \neq c_j$ when $i \neq j$. Each lot of $n_i$ balls of colour $c_i$ corresponds to the $n_i$ individuals whose native language is $L_i$.

After mixing the balls in the urn thoroughly, we draw two balls. This corresponds to the random selection of two individuals from $S_0$.

There are of course two ways of doing this: (i) a ball is drawn, its colour noted, and then after this ball is returned and the balls are mixed again, a second ball is drawn and its colour noted; or (ii) a ball is drawn, the remaining balls are mixed, and then a second ball is drawn and the colours of the two balls are noted. Scheme (i) is called (random) *sampling with replacement* and scheme (ii) is (random) *sampling without replacement*.

In (i), sampling with replacement, the set of possible ordered pairs of colours obtained from the experiment constitutes one of the possible sample spaces. Suppose, however, that each of the identical balls is inscribed with a number designating the individual in $S_0$ to which the ball corresponds, so that each ball is distinguished from every other one. Then a possible set of simple events to be considered is the set of ordered pairs of balls. This is essentially what we call the cartesian product $S_0 \times S_0$ of $S_0$ by $S_0$ as defined in (*l*) of section 0.1. Each ball in the urn has the same probability of being drawn in one draw, and, since we are replacing the ball after it is drawn and before the second ball is drawn, any pair of balls $(b_1, b_2)$ has the same probability of being drawn as any other:

(1)  $P(\{(b_1, b_2)\}) = 1/|S_0|^2 = 1/n^2.$

If we let $E$ be the set of all subsets of $S_0 \times S_0$ and define the probabilities of events as in exercise 1 of section 2.2, we find that $E$ and $P$ constitute a probability space.

Various events can be considered:
(i) the event

$$A_{i.} = \{(x, y) \mid x \text{ is a ball coloured } c_i\}$$

corresponding to the choice of an individual whose native language is $L_i$ upon the first try,
(ii) the event

$$A_{.j} = \{(x, y) \mid y \text{ is a ball coloured } c_j\}$$

corresponding to the choice of an individual whose native language is $L_j$ upon the second try,
(iii) the event

$$A_{ij} = \{(x, y) \mid x \text{ is coloured } c_i \text{ and } y \text{ coloured } c_j\}$$

corresponding to the choice of an individual whose native language is $L_i$ on the first try and another whose native language is $L_j$ on the second try.

If we select an individual ball $x_0$, and consider

$$B_{x_0} = \{(x, y) \mid x = x_0\} = \bigcup_{y \in S_0} \{(x_0, y)\},$$

we find that, since there are $n$ pairs in $B_{x_0}$ which, when taken as sets, are mutually exclusive,

$$P(B_{x_0}) = \sum_{y \in S_0} P\{(x_0, y)\} = \frac{|S_0|}{|S_0|^2} = \frac{n}{n^2} = \frac{1}{n}.$$

Similarly for any $i = 1, 2, \ldots, m$, the event $A_{i.}$ is the conjunction of all events $B_x$ for $x \in L_i$, i.e.,

$$A_{i.} = \bigcup \{B_x \mid x \text{ corresponds to a speaker of } L_i\}$$

where the $B_x$'s form a mutually exclusive set. Therefore since there are $n_i$ balls corresponding to speakers in $L_i$,

$$(2) \quad P(A_{i.}) = \sum_{\substack{x \text{ corresponds} \\ \text{to a speaker} \\ \text{of } L_i}} P(B_x) = \frac{n_i}{n}$$

for $i = 1, 2, \ldots, m$. We can also show that

$$(3) \quad P(A_{.j}) = n_j/n,$$

and that

$$(4) \quad P(A_{ij}) = n_i n_j/n^2$$

for $i, j = 1, 2, \ldots, m$. The events $A_{i.}$ and $A_{.j}$ are independent because

$$(5) \quad A_{ij} = A_{i.} \cap A_{.j},$$

and by (2), (3), and (4),

$$P(A_{i.} \cap A_{.j}) = P(A_{ij}) = \frac{n_i}{n}\frac{n_j}{n} = P(A_{i.})P(A_{.j}).$$

Since the collection of events

$$\{A_{ij} \mid i, j = 1, 2, ..., m\}$$

is a mutually exclusive set of events, in particular

$$(6) \quad P\left(\bigcup_{i=1}^{m} A_{ii}\right) = \sum_{i=1}^{m} P(A_{ii}) = \sum_{i=1}^{m} \frac{n_i^2}{n^2}.$$

The event $\bigcup_{i=1}^{m} A_{ii}$ can be described as the event: 'for a pair of balls $(x, y)$ selected at random, $x$ and $y$ have the same colour.' This event in the intermediate urn model corresponds to the event that a pair of speakers $(x, y)$ selected at random from $S_0 \times S_0$ speak the same language.

Since in the above discussion we are not interested in individual speakers, we could choose as an alternative sample space

$$M = \{(x, y) \mid x \text{ and } y = 1, 2, ..., m\}$$

where each number pair $(i, j)$ corresponds to the selection of a ball of colour $c_i$ in the first draw and one of colour $c_j$ in the second. Since there are $n_i$ out of $n$ balls coloured $c_i$ when the first draw is made and $n_j$ out of $n$ coloured $c_j$ when the second is made, it is natural to assign

$$P_0\{(i, j)\} = n_i n_j / n^2.$$

Now let us consider the second manner of selecting pairs of individuals from $S_0$, i.e., *selection without replacement*. Suppose as before the urn contains for each $i = 1, 2, ..., m$, $n_i$ balls of colour $c_i$ to correspond to the individuals in $S_0$ who have $L_i$ as their native language, and each ball is distinctively numbered. Assuming that the balls can be distinguished one from another, the population $S_1$ can be taken to be composed of the $n(n-1)$ pairs $(x, y)$ of balls where ball $x$ is distinct from ball $y$. Then we can take the set of possible outcomes of a 'draw' to be equal to $S_1$. Since each pair $(x, y)$ is as likely to be drawn as any other, the probability that a particular $(x, y)$ is drawn is $1/[n(n-1)]$. Using the techniques of exercise 1 in section 2.2, we can assign a probability $P_1(A)$ to every set $A$ in $E_1$, the set of all subsets of $S_1$, so that $E_1$ and $P_1$ again constitute a probability space.

Suppose we wish to know the probability that the first ball selected is coloured $c_i$ and the second $c_j$. The first ball can be selected in $n_i$ ways and the second in either $n_j$ ways, if $j \neq i$, or $n_i - 1$ ways, if $j = i$. Since there are $n(n-1)$

possible ways of selecting the two balls (without replacement), each as likely as any other,

$$(7) \quad P_{ij} = \begin{cases} \dfrac{n_i n_j}{n(n-1)} & \text{for } i \neq j. \\ \dfrac{n_i(n_i-1)}{n(n-1)} & \text{for } i = j. \end{cases}$$

is the probability that the first ball drawn is of colour $c_i$ and the second $c_j$. If we call this event $A_{ij}$ as in the replacement case, then the probability that two balls have the same colour is

$$(8) \quad P\left(\bigcup_{i=1}^{m} A_{ii}\right) = \sum_{i=1}^{m} P(A_{ii}) = \sum_{i=1}^{m} P_{ii} = \sum_{i=1}^{m} \frac{n_i(n_i-1)}{n(n-1)}.$$

Thus, if we sample with replacement, Greenberg's measure takes the form (6), and if we sample without replacement, it is given by expression (8).

Note that in sampling without replacement the probability that the first ball is a $c_i$ is

$$P(A_{i.}) = n_i/n$$

as in replacement sampling. However, when sampling *without* replacement, the computation of the probability of obtaining a ball of colour $c_j$ on the second draw is a little more complicated. By an extension of the results of equation (8) and exercise 11 in section 2.3, we can easily show that

$$(9) \quad P(A_{.j}) = \sum_{i=1}^{m} P(A_{.j}|A_{i.})P(A_{i.}).$$

Since for $i \neq j$

$$P(A_{.j}|A_{i.}) = n_j/(n-1)$$

and for $i = j$

$$P(A_{.j}|A_{j.}) = (n_j-1)/(n-1),$$

equation (9) becomes

$$P(A_{.j}) = \left(\sum_{i \neq j} \frac{n_j}{n-1}\frac{n_i}{n}\right) + \frac{n_j-1}{n-1}\frac{n_j}{n} = \frac{n_j}{n(n-1)}\left[\left(\sum_{i=1}^{m} n_i\right) - 1\right]$$

$$= \frac{n_j(n-1)}{n(n-1)} = \frac{n_j}{n},$$

which is also the case when sampling with replacement. From equation (7), the probability of drawing first a $c_i$ ball and then a $c_j$ ball is

$$P(A_{i.} \cap A_{.j}) = P(A_{ij}) = P_{ij} = \begin{cases} \dfrac{n_i(n_i-1)}{n(n-1)} & \text{if } i = j, \\[3ex] \dfrac{n_i n_j}{n(n-1)} & \text{if } i \neq j. \end{cases}$$

Since

$$P(A_{i.})P(A_{.j}) = n_i n_j/n^2 \neq P(A_{i.} \cap A_{.j}),$$

the two events $A_{i.}$ and $A_{.j}$ are not independent when the sampling is done without replacement.

If the various mother-tongue groups are large enough so that

$$\frac{n_i(n_i-1)}{n(n-1)} \quad \text{and} \quad \frac{n_i n_j}{n(n-1)}$$

are not significantly different from

$$n_i^2/n^2 \quad \text{and} \quad n_i n_j/n^2$$

respectively, then it is safe to assume independence because the effect of sampling without replacement is negligible.

EXAMPLE 2   In Ross (1950; see especially p. 27) the problem of genetic relationships among certain Indo-European languages is considered. His method consists of constructing a table as follows. Allot a column to each branch language and a row to each of a certain set of attested Indo-European roots; if a root appears in a particular branch language put an 'x' in the appropriate cell of the table:

| Root-number | $L_1$ | $L_2$ | $L_3$ | ... |
|---|---|---|---|---|
| 1 | x | x | | |
| 2 | | x | x | |
| 3 | x | x | x | |
| ⋮ | | | | |

He suggests that the question 'Is $L_i$ closely related to $L_j$?' is equivalent to the question 'Given the number of crosses in the $i$th and $j$th columns, what is the probability of obtaining the given number (or a greater number) of cases of a row with a cross in each of the two columns if the crosses were placed in the two columns at random?' If this probability is sufficiently small, we might be tempted to infer that the two languages have some causal (genetic) relationship.

Let $N$ be the number of rows in the table (number of roots under consideration), let $n_i$ be the number of rows which contain a cross in column $i$, let $n_j$ be the number of rows which contain a cross in column $j$, and let $r$ be the number of rows in which both the $i$th and $j$th columns are marked with a cross.

Our problem is to compute the probability that $r$ agreeing rows occur by chance, given that $n_i$ and $n_j$ entries in column $i$ and column $j$ respectively are marked with a cross.

We proceed as follows. First we determine the number of ways of marking $n_i$ entries in column $i$ and $n_j$ entries in column $j$ with exactly $r$ agreements. The number of ways that $n_i$ crosses can be placed in the $i$th column is

$$\binom{N}{n_i} = \frac{N!}{(n_i)!\,(N-n_i)!},$$

the number of ways of selecting a subset of $n_i$ objects from a set containing $N$ objects. (If this notation, first introduced in section 1.6, seems mysterious, work exercises 1 and 2 below and consult Brainerd et al., 1967, volume II, for further details.) Once these have been chosen, we put $r$ crosses in the $j$th column on $r$ of the $n_i$ rows already marked in column $i$. This can be done in $\binom{n_i}{r}$ ways. Finally we mark at random the $N-n_i$ rows which do not contain a cross in the $i$th column with the $n_j-r$ remaining crosses allotted to the $j$th column. This can be accomplished in

$$\binom{N-n_i}{n_j-r}$$

ways. Thus the entire operation can be done in

$$\binom{N}{n_i}\binom{n_i}{r}\binom{N-n_i}{n_j-r}$$

different ways.

Now we must find the number of ways $n_i$ crosses can be placed in column $i$ and $n_j$ in column $j$ without any restriction on agreements. Since $n_i$ of the entries of column $i$ can be marked with a cross in $\binom{N}{n_i}$ ways and $n_j$ of the entries of column $j$ can be marked with a cross in $\binom{N}{n_j}$ ways, the total number of ways both columns can be marked is

$$\binom{N}{n_i}\binom{N}{n_j}.$$

We let the set of outcomes be the set of all these possible ways of marking the two columns with crosses. Since each way is as likely to occur as any other, the probability of any one such marking is

$$\frac{1}{\binom{N}{n_i}\binom{N}{n_j}},$$

and since, as we have seen,

$$\binom{N}{n_i} \binom{n_i}{r} \binom{N-n_i}{n_j-r}$$

cases are favourable to the event of $r$ agreeing markings, the probability $p_r$ of $r$ agreements is

$$p_r = \binom{N}{n_i} \binom{n_i}{r} \binom{N-n_i}{n_j-r} \bigg/ \binom{N}{n_i} \binom{N}{n_j}$$

or

$$(10) \quad p_r = \binom{n_i}{r} \binom{N-n_i}{n_j-r} \bigg/ \binom{N}{n_j}.$$

Suppose that we find $s$ coincident crosses when we consider the actual case of Indo-European roots in $L_i$ and $L_j$. The probability of at least that many occurring by chance is

$$P(m \geq s) = \sum_{q=s}^{k} p_q$$

where $k$ is the smaller of the two numbers $n_i$ and $n_j$. If $P(m \geq s)$ is small, say 0.05, this means that the probability of obtaining at least as many as $s$ coincident roots by chance is 0.05. Since the probability of obtaining $s$ or more common roots by chance is small indeed, one is tempted to reject the hypothesis of random root coincidence and posit some non-random mechanism which accounts for the phenomenon. Thus it would appear that some causal factor is at work. If, on the other hand, $P(m \geq s)$ turned out to be, say, 0.5, we would not be tempted to look for causal factors, for in this case there is a 50–50 chance that $s$ or more roots coincide. This sort of argument is basic to hypothesis testing, which is the subject of chapter 5, to follow.

Ross's problem can also be considered in terms of an urn model. Assume that $n_i \geq n_j$. Let an urn be filled with $N$ balls of which $n_i$ are red and the rest $(N-n_i)$ are white. The $n_i$ red balls correspond to the $n_i$ crosses in the $i$th column. A sample of $n_j$ balls is drawn from the urn. This corresponds to the random marking of the $j$th column with $n_j$ crosses. What is the probability that $r$ of this sample of $n_j$ balls are red? Or, in other words, what is the probability of obtaining $r$ coincident crosses?

A sample of $n_j$ balls containing $r$ red balls and $n_j-r$ white balls can be drawn from the urn in

$$\binom{n_i}{r} \binom{N-n_i}{n_j-r}$$

ways, and there are $\binom{N}{n_j}$ ways of drawing samples of $n_j$ balls from the urn. Since these are equally likely,

$$(10) \; p_r = \binom{n_i}{r} \binom{N-n_i}{n_j-r} \Big/ \binom{N}{n_j} .$$

Some readers may observe that the original problem is entirely symmetric with respect to $i$ and $j$ so that it should be possible to show that we can also write

$$(10') \; p_r = \binom{n_j}{r} \binom{N-n_j}{n_i-r} \Big/ \binom{N}{n_i} .$$

The reader is invited to prove this in exercise 2(d) below. We shall return shortly to the $p_r$'s defined by equation (10). The set $\{p_0, p_1, ..., p_{n_i}\}$ in (10) constitutes a particular case of the *hypergeometric distribution* which has been widely studied.

Ross's model is mathematically identical with that assumed in Chrétien (1943), as we shall see in chapter 5. His procedure is actually a standard technique in the analysis of binary data. See Cox (1970, §4.3).

EXERCISES

1 A well-known rule of thumb for counting the number of ways of selecting pairs of objects is the following. If an object $o_1$ can be selected in $m$ ways and an object $o_2$ can be selected in $n$ ways, then the ordered pair of objects $(o_1, o_2)$ can be selected in $mn$ ways.

(a) Use this rule to show that an ordered pair of balls can be selected (without replacement) from an urn containing $n$ distinctly marked balls in $n(n-1)$ distinct ways.

(b) If the order of selection of the balls in (a) is of no importance, show that the number of ways of selecting the two balls is $n(n-1)/2$.

(c) Show that if three balls are involved instead of two in questions (a) and (b), then the answers to these questions are respectively $n(n-1)(n-2)$ and $n(n-1)(n-2)/3\cdot2$.

(d) We write $k!$ ($k$ *factorial*) as the product

$$k! = k\cdot(k-1)\cdot(k-2)\cdot...\cdot2\cdot1$$

when $k \geqq 1$ and define $0! = 1$. Using this notation show that

(i) $\quad n(n-1) = \dfrac{n!}{(n-2)!},$

(ii) $\quad \dfrac{n(n-1)}{2!} = \dfrac{n!}{(n-2)!\,2!},$

(iii) $\quad \dfrac{n(n-1)(n-2)}{3\cdot2} = \dfrac{n!}{(n-3)!\,3!}.$

(e) Show that the number of ways of choosing an ordered set of $k$ balls (without replacement) from an urn containing $n$ distinctly marked balls is $n!/(n-k)!$

(f) And if we have no interest in the order in which these balls are drawn, i.e., if we require only the number of ways of selecting a set of $k$ balls from the urn, then show that the answer is $n!/[(n-k)!\,k!]$. This number, i.e., the number of ways that a set of $k$ distinct objects can be selected from a set of $n$ distinct objects, is usually symbolized by

$$\binom{n}{k} = \frac{n!}{(n-k)!\,k!}$$

and is sometimes called the number of *combinations of n things taken k at a time* (or, colloquially, *n choose k*). If $k > n$ or $k < 0$, we adopt the convention that $\binom{n}{k} = 0$.

2 Show that the following equations hold:

(a) $\binom{n}{k} = \binom{n}{n-k}$,

(b) $\binom{n}{k} = \binom{n-1}{k-1} + \binom{n-1}{k}$,

(c) $\binom{n}{0} + \binom{n}{1} + \ldots + \binom{n}{n} = 2^n$

(remember that $0! = 1$ by definition, and use the binomial theorem [see exercise 1 in section 1.6]),

(d) $\dfrac{\dbinom{n}{r}\dbinom{N-n}{k-r}}{\dbinom{N}{k}} = \dfrac{\dbinom{k}{r}\dbinom{N-k}{n-r}}{\dbinom{N}{n}}$    for $N \geq n$, $N \geq k$ and $r \leq n$, $r \leq k$.

3 In a game of dice, what is the probability of throwing each of the various points? The answer can be obtained by following the procedure laid out below in (a)–(f). Distinguish the two dice and let the outcome of a toss be $(x, y)$, where $x$ is the number on the upper face of die I and $y$ the number on die II.

(a) Show that the set of all possible outcomes is

$U = \{(x, y) \mid x, y = 1, 2, \ldots, 6\}$.

(b) If all the outcomes in $U$ are deemed equally likely, show that for $A \subseteq U$ the appropriate probability is $P(A) = |A|/36$, where of course $|A|$ is the number of elements in $A$.

(c) Prove that $E, P$ is a probability space if $E$ is taken to be the set of all subsets of $U$.

(d) Show that the 'point' $z$ corresponds to the event

$E_z = \{(x, y) \in S \mid z = x+y\}$.

(e) Find $E_z$ for $z = 1, 2, 3, 4, 5, \ldots, 12, 13, \ldots$

(f) Find the probability of $E_z$ for each $z$.

(g) Find the event $H_x$ corresponding to $x$ on the first die for $x = 1, 2, \ldots, 6$.

(h) Find the event $G_y$ corresponding to $y$ on the second die for $y = 1, 2, \ldots, 6$.

(i) Show that $H_x$ and $G_y$ constitute an independent pair of events for any choice of $x$ and $y$ between 1 and 6 inclusive.

4 Using an urn model, attempt to justify the statement made in section 1.1 that a simple sample and a random sample (with replacement) are the same thing.

D

# 3
# Random variables

RANDOM VARIABLES AND THE DEFINITE INTEGRAL

### 3.1.1 *The notion of random variable*

Earlier we encountered the notion of a variate, examples of which were

$$W = \frac{\text{no. of noun-tokens in a paragraph}}{\text{no. of word-tokens in the paragraph}}$$

in section 1.1,

$Y = $ no. of different Chinese characters in 50 characters taken from the *Tao Teh Ching*

in example 1 of section 1.1, and

$Z = $ no. of articles in a 50-word passage

in example 2 of section 1.6. These and nearly all the other variates we have considered and shall consider in this book have certain properties definable in terms of probability.

Take for example $Z$, the number of articles in a 50-word passage of some text. The set of possible outcomes, or values of $Z$, when a 50-word passage is chosen at random is, taking the most liberal view possible, $\{0, 1, 2, \ldots, 50\}$. Later we shall make some hypotheses concerning the probabilities of these outcomes, so that for example the events $Z = 1$, $Z \leqq 3$, $0 < Z$, etc., can all be assigned a probability. In cases like this we say the variate is a 'random variable.' This concept is very basic to probability and statistics. It is neither random nor a variable but is, rather, a special kind of real-valued function defined over the universal set $U$ of all possible outcomes of some experiment. If the probability space corresponding to this experiment is composed of the set of events $E$ and the probability $P$, then a *random variable over* that probability space is any function $X$ from $U$ into the set of real numbers such that the set

$$(X \leqq x) = \{u \in U \mid X(u) \leqq x\}$$

belongs to $E$, i.e., is an event, for all choices of the real number $x$.

This definition ensures that every event of the form $(X \leqq x)$, i.e., every event of the form

'the variate $X$ has a value less than or equal to $x$,'

has a known probability. From this definition and the properties of a (continuous) probability space, we can show that it is possible to compute the probabilities corresponding to many other events defined in terms of $X$, e.g.,

$$(a < X \leqq b) = \{u \in U \mid a < X(u) \leqq b\}, \quad (X = a) = \{u \in U \mid X(u) = a\},$$

$$(a < X < b) = \{u \in U \mid a < X(u) < b\},$$

$$(a < X) = \{u \in U \mid a < X(u)\}, \quad (a \leqq X) = \{u \in U \mid a \leqq X(u)\}.$$

In section 1.1 we introduced the idea of a variate $X$ as a function defined on a population. In section 2.2 we noted that a possible interpretation of the concept of experiment introduced there was the selection of an individual at random from a population and the measurement of the value of some variate $X$ for that individual. If it is possible to define a probability space $E$, $P$ corresponding to that experiment where the collection of events $(X \leqq \lambda)$ belongs to $E$ for all real numbers $\lambda$, then of course the variate $X$ becomes a random variable. In the sequel, we shall consider only variates that are random variables and shall use the terms interchangeably as long as no confusion can arise.

We have already encountered random variables, for example when we discussed the urn model in connection with example 2 of section 2.4. There we considered an urn containing two kinds of identically shaped balls, say $k$ red balls and $N-k$ white balls. We withdrew $m$ balls from the urn and asked the question, 'What is the probability that $r$ of the $m$ balls drawn are red?' As we saw in example 1 of section 2.4 (when $m = 2$), one way of looking at this problem is to assume that each ball in the urn can be distinguished from every other one and then let $U$ be the collection of $\binom{N}{m}$ different possible sets

$$s = \{b_1, b_2, ..., b_m\}$$

of $m$ different balls which can be drawn (without replacement) from the urn. If we let $E$ be the set of all subsets of $U$, then because each set $s = \{b_1, ..., b_m\}$ is as likely to be drawn as any other, the probability of drawing a particular set $s$ is

$$1 \Big/ \binom{N}{m}.$$

An intuitively natural probability can be assigned to each of the events $A$ in $E$ by letting

$$P(A) = (\text{no. of elements of } A) \Big/ \binom{N}{m}.$$

The function $R$ defined on $U$ by the expression '$R(s)$ = number of red balls in $s$' can be seen to be a random variable. Indeed, for each real number $r$ the event $(R \leq r)$ belongs to $E$ and hence has a probability. To find this probability, note that for $0 \leq r \leq m$

$$P(R \leq r) = \sum_{i \leq r} P(R = i)$$

$$= \sum_{i \leq r} \{\text{probability that the sample contains } i \text{ red balls}\}.$$

From the results of section 2.4, we conclude that

$$(1) \quad P(R \leq r) = \sum_{i \leq r} \binom{k}{i} \binom{N-k}{m-i} \Big/ \binom{N}{m}.$$

Thus if a sample of $m$ objects is drawn without replacement from a population of $N$ objects among which $k$ possess a certain property, and if $R$ is the number of individual objects in the sample possessing this property, then the probability that $R = i$ is given by the expression

$$(2) \quad P(R = i) = h(i; m, k, N) = \binom{k}{i} \binom{N-k}{m-i} \Big/ \binom{N}{m}.$$

$R$ in this case is said to have the *hypergeometric distribution with parameters $m$, $k$, $N$*. We use $h(i; m, k, N)$ to stand for the $i$th value of the hypergeometric distribution. The parameters $m$, $k$, $N$ determine which of the hypergeometric distributions we are using; there is, of course, a different hypergeometric distribution for each of the possible choices of $m$, $k$, and $N$.

Since $R(s)$ can equal only $0, 1, 2, \ldots, m$, $R < 0$ is impossible and $R \leq m$ is certain to be true. Therefore $P(R \leq r) = 0$ for $r < 0$ and $P(R \leq r) = 1$ for $r \geq m$.

A graph of $P(R \leq r) = F_R(r)$ for specific values of $N$, $k$, and $m$, say $N = 8$, $k = 3$, and $m = 4$, might prove instructive. The results for this case are displayed in table 3.1. To obtain the various values of the $F_R(r)$ given below, we use this table together with expression (2):

$$(3) \quad F_R(r) = \begin{cases} 0 & \text{for } r < 0, \\ 1/14 & \text{for } 0 \leq r < 1, \\ 1/2 & \text{for } 1 \leq r < 2, \\ 13/14 & \text{for } 2 \leq r < 3, \\ 1 & \text{for } 3 \leq r. \end{cases}$$

The graph of $F_R(r)$ is depicted in figure 3.1.

An intuitively compelling way to consider the probabilities connected with the random variable $R$ is to think of the probability that $R(s) = r$ as a mass of

TABLE 3.1

| $i$ | $P(R = i)$ |
|---|---|
| 0 | 1/14 |
| 1 | 3/7 |
| 2 | 3/7 |
| 3 | 1/14 |
| 4 | 0 |

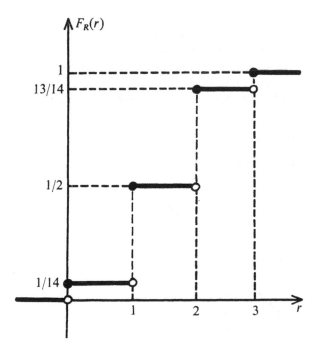

FIGURE 3.1

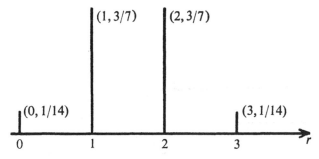

FIGURE 3.2  Each vertical line represents a mass equal to its length concentrated at the corresponding point on the $r$-axis.

material distributed along the real line. Thus the information in table 3.1 can be presented as in figure 3.2, using this analogy. $F_R(r)$ is then the amount of mass at or to the left of $r$.

Usually in linguistic applications we encounter finite or countably infinite (discrete) populations. However, on certain occasions continuous populations must be considered. The following example provides the simplest such population.

EXAMPLE 1 (uniform distribution on [0, 1])   In example 3 of section 2.2 we considered the following experiment. A point is selected at random from the interval $[0, 1] = \{x \mid 0 \leq x \leq 1\}$. Since every point is as likely to be drawn as any other and since there are more than a finite number of these points, we cannot assign to the probability of drawing a particular point the value one over the number of points, and hope to obtain from this the probability of drawing a point in, say, the set

$$[\alpha, \beta] = \{x \in [0, 1] \mid \alpha \leq x \leq \beta\}$$

for $\alpha < \beta$. However, in section 2.2 we considered another intuitively natural way

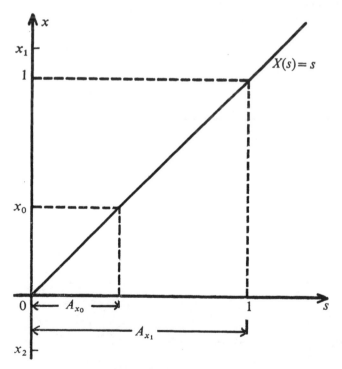

FIGURE 3.3   $A_{x_0} = \{s \in [0,1] \mid X(s) \leq x_0\} = [0, x_0], A_{x_1} =$ $\{s \in [0,1] \mid X(s) \leq x_1\} = [0,1] = S, A_{x_2} = \{s \in [0,1] \mid X(s) \leq x_2\} = \emptyset$.

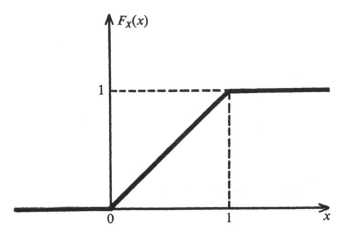

FIGURE 3.4

of defining the probability space connected with this experiment. We let the set of outcomes be all points in [0, 1]. Let $E$ be the smallest collection of subsets of [0, 1] which contains all the intervals of the form [$\alpha$, $\beta$] for $0 \leqq \alpha < \beta \leqq 1$ and satisfies axioms of a continuous probability space. (This choice of $E$ is a large but *proper* subset of $2^{[0,1]}$ whose use is dictated by technical considerations which we need to go into here [see Burkhill 1961]. This $E$ contains, in addition to intervals, one-point sets, the set of rational numbers in [0, 1], the set of irrational numbers in [0, 1], and so on. Each of these sets can be assigned a probability.) Finally we let the probability that the point selected lies in [$\alpha$, $\beta$] be the ratio of the length of [$\alpha$, $\beta$] to the length of [0, 1], i.e.,

(4)  $P([\alpha, \beta]) = \dfrac{\beta - \alpha}{1 - 0} = \beta - \alpha.$

Now consider the random variable $X$, where

(5)  $X(s) = s,$

that assigns to each $s \in [0, 1]$ the distance of $s$ from the origin. Then

(6)  $P(X \leqq x) = P(\{s \in [0, 1] \mid X(s) \leqq x\}) = P([0, x]) = \dfrac{x - 0}{1 - 0} = x$

for $0 < x \leqq 1$, and $P(X \leqq x)$ equals 0 for $x \leqq 0$ and 1 for $x > 1$ (see figure 3.3). The graph of $F_X(x) = P(X \leqq x)$ is then given in figure 3.4.

If we view the probability of the various values of $X$ as a distribution of mass along the line, then, instead of discrete points of fixed but positive mass (probability), we find the mass (probability) uniformly distributed on the interval

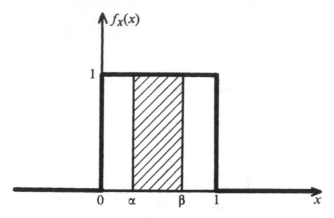

FIGURE 3.5

[0, 1], as indicated in figure 3.5, so that the mass between $\alpha$ and $\beta$ is exactly the area of the shaded region. The function $f_X$ depicted in figure 3.5 is defined by

$$(7) \quad f_X(x) = \begin{cases} 0 & \text{for } x < 0, \\ 1 & \text{for } 0 \leq x \leq 1, \\ 0 & \text{for } x > 1. \end{cases}$$

To the reader with training in the calculus, it is clear that

$$(8) \quad F_X(x_0) = \int_{-\infty}^{x_0} f_X(x)dx = \begin{cases} 0 & \text{for } x_0 < 0, \\ x_0 & \text{for } 0 \leq x_0 \leq 1, \\ 1 & \text{for } 1 < x_0. \end{cases}$$

For those not familiar with the calculus, this formula can be construed as giving the area to the left of the line $x = x_0$ and between the line $y = f_X(x)$ and the $x$-axis.

### 3.1.2 *The definite integral*

For those without prior knowledge of the calculus, the following discussion will explain the symbolism of (8). If $f$ is a continuous function from the real numbers to the real numbers whose values are never negative, then the expression

$$(9) \quad \int_a^b f(x)\, dx$$

stands for *the (definite) integral of $f(x)$ from $a$ to $b$* and is equal to the area of the region bounded by the lines with equations $x = a$, $x = b$, $y = 0$ (the horizontal axis), and $y = f(x)$. The shaded area in figure 3.6 is such a region.

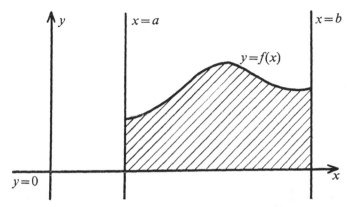

Figure 3.6   The area of the shaded region is designated $\int_a^b f(x)dx$.

If $a$ is not a number but $-\infty$ ('minus infinity'), then

$$\int_{-\infty}^b f(x)\,dx,$$

the integral of $f(x)$ from minus infinity to $b$, represents the area of the (possibly) infinitely extensive region between $y = f(x)$ and $y = 0$ and to the left of $x = b$, as in figure 3.7, for example.

In expression (9), $a$ is the *lower limit of integration*, $b$ the *upper limit of integration*, $x$ the *variable of integration*, and the function $f(x)$ is the *integrand*.

The integral calculus provides us with techniques for calculating integrals (i.e., areas)

$$\int_a^b f(x)\,dx$$

when $a$, $b$, and $f(x)$ are known. For a large number of $f$'s important to probability and statistics, the values of the integral are tabulated in terms of $a$ and $b$, so that *it is not necessary to know the calculus in order to utilize and understand this notation in statistical problems.*

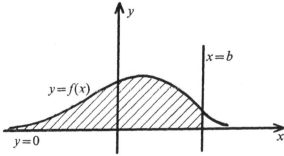

Figure 3.7   The area of the shaded region is designated $\int_{-\infty}^b f(x)dx$.

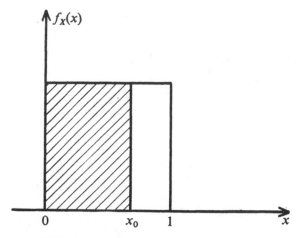

FIGURE 3.8

In expression (8), $F_X(x_0)$ stands for the area to the left of $x = x_0$ which is bounded by the line $y = 0$ and the curve $y = f_X(x)$ in expression (7) as shown in figure 3.8. Note that for $x \leq 0$ the two curves $y = 0$ and $y = f_X(x)$ coincide, and hence the area between them and to the left of $x_0$ is 0 so long as $x_0 \leq 0$. However if $0 < x_0 \leq 1$ our knowledge of geometry tells us that

$$F_X(x_0) = x_0.$$

For $x > 1$, $f_X(x) = 0$, so again for such values of $x$ the curves $y = 0$ and $y = f_X(x)$ coincide. Thus no more area is added after $x = 1$, and so

$$F_X(x_0) = 1$$

for $x_0 \geq 1$; thence $F_X$ takes the form indicated in figure 3.4.

In some cases the value of the area under a curve and to the left of some vertical line $x = x_0$ is not so easily arrived at, as the following example indicates.

EXAMPLE 2   The function

$$(10) \; f(x) = \frac{e^{-x^2/2}}{\sqrt{2\pi}} \quad (-\infty < x < \infty),$$

where $e = 2.71828\ldots$, characterizes a fundamentally useful function in probability and statistics. The area bounded by the $x$-axis $(y = 0)$, $y = f(x)$ and to the left of a line $x = \xi$, i.e., the integral

$$(11) \; \int_{-\infty}^{\xi} \frac{e^{-x^2/2}}{\sqrt{2\pi}} \, dx = P(\xi$$

is widely tabulated (see, for example, Pearson and Hartley [1962], p. 1 and

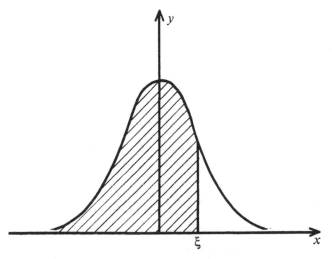

FIGURE 3.9

table 1). The graph of $y = f(x)$ is given in figure 3.9, where for each $x, f(x) > 0$ although as $x \rightarrow \pm \infty$, $f(x) \rightarrow 0$. Expression (11) represents the area of the shaded region in figure 3.9.

Usually the value of (11) is given for positive $\xi$'s. We shall see shortly how the area represented by the integral

$$\int_a^b \frac{e^{-x^2/2}}{\sqrt{2\pi}} \, dx$$

can be obtained for arbitrary values of $a < b$. The function defined by (10) is called the *unit normal density function*.

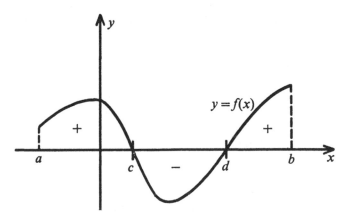

FIGURE 3.10

In general, if $f(x)$ is not always positive, as in figure 3.10, then when the graph of $f(x)$ dips below the $x$-axis, the area between it and the $x$-axis is counted negatively Thus for the $f(x)$ in figure 3.10,

$$\int_a^b f(x) \, dx = \text{(area under } f \text{ between } a \text{ and } c)$$

$$- \text{(area 'under' } f \text{ between } c \text{ and } d)$$

$$+ \text{(area under } f \text{ between } d \text{ and } b).$$

The remarks of this subsection should enable readers without prior knowledge of the calculus to interpret most of the uses we put it to in what follows.

### 3.1.3 *The distribution of a random variable*

Using the analogy of probability to mass distributed along a line, we note that in both the discrete case (the urn example in the first part of the section) and the continuous case (example 1) we defined a function which gave for each real number the probability that the random variable in question had a value less than or equal to that real number. In the urn example this function was

(12) $F_R(r) = P(R \leqq r)$,

given by expression (3), and in example 1 it was

(13) $F_X(x) = p(X \leqq x)$,

given by expression (8).

In general, if $Y$ is a random variable on some probability space, then the *cumulative distribution function* (CDF) *of $Y$* is defined to be the function

$$F_Y(x) = P(Y \leqq x).$$

Expression (1) is a cumulative distribution function that is best expressed as a sum because $R$ takes on only a discrete set of values with non-zero probability. Thus $R$ is said to be *discretely* distributed. The CDF given by (8), however, is expressed as an integral because $P(X \leqq x)$ varies continuously with $x$ and so we say that $X$ is *continuously* distributed.

The integral

$$\int_a^b f(x) \, dx$$

is a continuous analogue of the (discrete) sum

$$\sum_{i=m}^n f_i$$

of $f_i$ from $m$ to $n$. In fact, the integral shares many properties with the sum. Among these the following are most important:

(14) $\int_a^b cf(x)\,dx = c\int_a^b f(x)\,dx,$

i.e., the integral (from $a$ to $b$) of a constant $c$ times $f(x)$ is equal to $c$ times the integral of $f(x)$;

(15) $\int_a^b [f(x)+g(x)]\,dx = \int_a^b f(x)\,dx + \int_a^b g(x)\,dx,$

i.e., the integral of the sum of two functions is the sum of the integrals of each of these functions evaluated separately;

(16) $\int_a^b f(x)\,dx = \int_a^b f(t)\,dt,$

i.e., the name of the variable of integration, like the name of the index of summation in a sum, does not affect the value of the integral; and finally, if $a < b < c$, then

(17) $\int_a^c f(x)\,dx = \int_a^b f(x)\,dx + \int_b^c f(x)\,dx.$

Expressions (14), (15), (16), and (17) are the analogues for integrals of the results of exercises 2(b), 2(a), equation (1), and exercise 3(c) respectively in section 1.2.

One additional observation will aid in understanding the notation. If $b$ approaches $c$ from the left, then

(18) $\lim_{b \to c} \int_a^b f(x)\,dx = \int_a^c f(x)\,dx.$

Thus if $X$ is continuously distributed so that

$$P(X \leq x) = \int_{-\infty}^x f(t)\,dt,$$

then from (18) we observe that for $z < x$

$$P(X < x) = \lim_{z \to x} P(X \leq z) = \lim_{z \to x} \int_{-\infty}^z f(t)\,dt = \int_{-\infty}^x f(t)\,dt = P(X \leq x).$$

Therefore for a continuously distributed random variable $X$,

$$P(X = x) = \lim_{z \to x} [P(X \leq x) - P(X \leq z)] = \lim_{z \to x} \left( \int_{-\infty}^x f(t)\,dt - \int_{-\infty}^z f(t)\,dt \right)$$

$$= \lim_{z \to x} \int_z^x f(t)\,dt = 0$$

for all $x$ (see figure 3.11). It should be remarked that more advanced texts also treat random variables which are a sum of continuous and discrete random variables. In our discussion, however, the two concepts can be kept separate.

The following two examples should help the reader to gain a fuller intuitive appreciation of the concepts of random variable and cumulative distribution

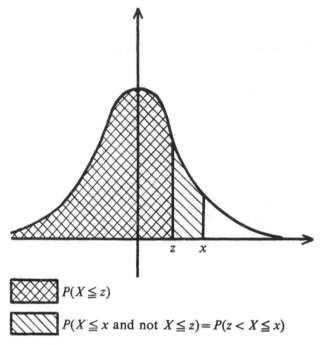

$P(X \leqq z)$

$P(X \leqq x$ and not $X \leqq z) = P(z < X \leqq x)$

FIGURE 3.11   As $z$ approaches $x$ from below, the area of the
singly hatched region approaches zero.

function for both the discrete and continuous cases, as well as introducing two
very important distributions.

EXAMPLE 3   In Roberts (1965, appendix II, p. 94), we find the relative frequency
of /ž/ given as 0.0003. Taking this as a probability, we might ask the question:
In a block of 1000 successive phonemes, what is the probability of $n$ occurrences
of /ž/? If we assume that the occurrence of a /ž/ in a particular position of the
block is independent of its occurrence in the next position (an assumption which
is strictly speaking not true but probably is approximately true in the sense that
the deviation from independence is not significant), then the above problem is
an example of a widely considered experimental situation.

Suppose we perform an experiment $\mathscr{E}$ with two outcomes, *success* or *failure* (in
example 3 the occurrence of /ž/ and the occurrence of another phoneme), such
that the probability of *success* is $p$ (0.0003 in example 3). Suppose also that this
experiment can be performed an arbitrary number of times, say $n$ (the 1000 in
example 3), such that the outcome in each trial is independent of the outcomes in
all the other trials and that $p$ remains constant over the trials. Such a sequence of
trials is called a sequence of *Bernoulli trials with probability $p$ of success*. The

probability of, say, $m$ successes followed by $n-m$ failures is

(20) $p^m(1-p)^{n-m}$

because the successive events are independent and each of the $m$ successes has probability $p$ while each of the $n-m$ failures has probability $1-p$. This event is only one of the $\binom{n}{m}$ events with $m$ successes and $n-m$ failures, each of which will have a probability given by (20). Since these events are mutually exclusive, the probability of the event, exactly $m$ successes, in a conjunction of $\binom{n}{m}$ events each with probability $p^m (1-p)^{n-m}$. Thus if $X$ stands for the number of successes in $n$ Bernoulli trials, then

(21) $P(X = m) = \binom{n}{m} p^m (1-p)^{n-m}.$

In general, any random variable $X$ satisfying (21) for $m = 0, 1, 2, \dots, n$ it is said to have a *binomial distribution with parameters n and p*. The *binomial weighting function*

(22) $b(k; n, p) = \binom{n}{k} p^k(1-p)^{n-k}$

determines the distribution of $X$ completely, for the CDF of $X$ can be written

$$F_X(x) = \sum_{k \leq x} b(k; n, p).$$

EXAMPLE 3 (continued)   Now assuming (i) that Roberts's relative frequency is near the probability of /ž/ in a given random selection of phonemes and (ii) that the textual occurrences of /ž/ and not-/ž/ are independent, then the text can be thought of as a set of Bernoulli trials, and so

(23) $P(Z = k) = \binom{1000}{k} (0.0003)^k (0.09997)^{1000-k}$

where $Z$ is the random variable with value

$Z(s) = $ number /ž/'s in a selection $s$ of 1000 phonemes.

Thus

$$P(Z = 0) = \binom{1000}{0} (0.0003)^0 (0.9997)^{1000}$$

$$= \frac{1000!}{0!1000!} (0.9997)^{1000} = (0.9997)^{1000}.$$

To compute $(0.9997)^{1000}$ we need logarithms. For those readers who have no prior knowledge of logarithms the following will be helpful. First, by definition, the *logarithm to the base b of a number a is the power to which b must be raised in order to equal a,* in formula

$$a = b^{\log_b a}.$$

Another approach is to say that

$$a = b^y \text{ means the same as } y = \log_b a.$$

Logarithms for $b = 10$ and $b = e$, where

$$e = \sum_{n=0}^{\infty} \frac{1}{n!} = 1 + 1 + \frac{1}{2} + \frac{1}{3!} + \frac{1}{4!} + \dots = 2.71828 \dots$$

have been extensively tabulated. Logarithms to the base $e$ are often called Naperian logarithms in honour of their inventor, John Napier (1550–1617). The main advantages of logarithms lie in their satisfying the following relations:

$$\log_b(xy) = \log_b x + \log_b y,$$

$$\log_b(x^q) = q \log_b x,$$

$$\log_b(x/y) = \log_b x - \log_b y.$$

Thus multiplications and divisions can be reduced to additions and subtractions. In the case of $(0.9997)^{1000}$ we have (for $b = 10$)

$$\log_{10}(0.9997)^{1000} = 1000 \log_{10}(0.9997) = 1000(9.999870 - 10)$$

$$= (9999.87000 - 10,000) = 9.87000 - 10.$$

So

$$(0.9997)^{1000} = \text{the number whose } \log_{10} \text{ is } 9.87000 - 10 = 0.7413.$$

Most desk-top computers have the log function built in so that the procedure of finding $(0.9997)^{1000}$ would go something like the following. Take the log of 0.9997, multiply the result by 1000, and take the anti-log of this second result. The final result will be $(0.9997)^{1000}$.

For higher values of $k$, the computation of $P(Z = k)$ in (23) becomes somewhat more arduous. We shall find later (in section 3.5) that in situations such as this $b(k; n, p)$ can be approximated by a much simpler expression in terms of the normal distribution.

### 3.1.4 *The normal distribution*

The values of certain random variables such as, for example, the height or weight of adult humans, scores on tests, and perhaps the Naperian logarithm of the length of a sentence in words (see Williams [1970] and papers by Williams and Buch in Doležel and Bailey [1969], pp. 69–75 and 76–9) follow the so-called normal distribution. This distribution is fundamental to statistics because it is the limiting distribution (as the sample size $n \to \infty$) of a large class of distributions, as we shall see in section 3.5.

A random variable is said to be *normally distributed* with *mean* $\mu$ and *variance* $\sigma^2$ if

(24) $P(X \leq x) = F_X(x) = N(x; \mu, \sigma)$

$$= \frac{1}{\sigma\sqrt{2\pi}} \int_{-\infty}^{x} e^{-(t-\mu)^2/2\sigma^2} dt.$$

The reader may have noticed that the terms mean and variance have already appeared in chapter 1, in each case prefaced by the word 'sample.' This is no coincidence, for we shall see later that if $n$ values $x_1, x_2, \ldots, x_n$ of a normally distributed random variable $X$ are sampled, then as the sample size $n$ increases, the sample mean

$$\bar{x} = \frac{1}{n} \sum_{i=1}^{n} x_i$$

and the sample variance

$$s^2 = \frac{1}{n} \sum_{i=1}^{n} (x_i - \bar{x})^2$$

approach respectively $\mu$, the mean of the distribution of $X$, and $\sigma^2$, the variance of the distribution of $X$. It is also noteworthy that once we know, by some means or other, that a random variable is normally distributed with mean $\mu$ and variance $\sigma^2$, our interest in the probability space on which it was defined diminishes, for we can obtain all the statistical properties of $X$ from expression (24). This remark clearly holds for other distributions as well. Thus *if we know the form of the distribution of a random variable X, we can obtain the statistical properties of X without reference to the probability space upon which it was defined.*

The integrand

(25) $\varphi(x) = \dfrac{e^{-(x-\mu)^2/2\sigma^2}}{\sigma\sqrt{2\pi}}$

in expression (24) is called the *probability density function of the normal distribution with mean $\mu$ and variance $\sigma^2$*. Its graph is the familiar bell-shaped normal curve. In figure 3.12 the graph of the normal density function is given when $\sigma = 1$ and $\mu$ takes on various values. In figure 3.13 the mean $\mu$ is fixed at zero and $\sigma$ takes on various values.

The reader will observe from figures 3.12 and 3.13 and expression (25) that the graph of the density function of a normally distributed random variable is

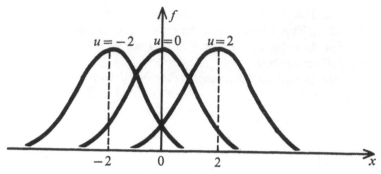

FIGURE 3.12

symmetric about its mean, i.e., about the vertical line $x = \mu$ where $\mu$ is the mean of the distribution. Thus the area under the graph of the normal density function in (25) and to the left of $\mu-a$ for $a > 0$ is equal to the area under the curve to the right of $\mu+a$. Therefore,

$$(26) \quad N(\mu-a; \mu, \sigma) = \int_{-\infty}^{\mu-a} \varphi(x)\, dx = \int_{\mu+a}^{+\infty} \varphi(x)\, dx$$

$$= 1 - \int_{-\infty}^{\mu+a} \varphi(x)\, dx = 1 - N(\mu+a; \mu, \sigma).$$

Further

$$(27) \quad P(|X-\mu| \leq a) = P(\mu-a \leq X \leq \mu+a) = \int_{\mu-a}^{\mu+a} \varphi(x)\, dx$$

$$= N(\mu+a; \mu, \sigma) - N(\mu-a; \mu, \sigma)$$

$$= N(\mu+a; \mu, \sigma) - [1 - N(\mu+a; \mu, \sigma)]$$

$$= 2N(\mu+a, \mu, \sigma) - 1$$

by (26). Since $|X-\mu|$ is either $\leq a$ or $> a$ but not both,

$$(28) \quad P(|X-\mu| \leq a) + P(|X-\mu| > a) = 1.$$

From (27), we immediately have

$$(29) \quad P(|x-\mu| > a) = 1 - P(|X-\mu| \leq a) = 2[1 - N(\mu+a; \mu, \sigma)].$$

The cumulative normal distribution

$$(30) \quad N(x; 0, 1) = \int_{-\infty}^{x} \frac{e^{-t^2/2}}{\sqrt{2\pi}}\, dt,$$

with $\mu = 0$ and $\sigma = 1$, as we have observed in example 2, is widely tabulated (Pearson and Hartley 1962) for values of $x \geq 0$. In a later section, we shall see how to obtain $N(x; \mu, \sigma)$ for arbitrary values of $\mu$ and $\sigma$ from these tables of (30).

However, as an exercise let us find $P(0 \leq X \leq 1)$ when $X$ is normally distributed with mean $\mu = 0$ and variance $\sigma^2 = 1$. Since

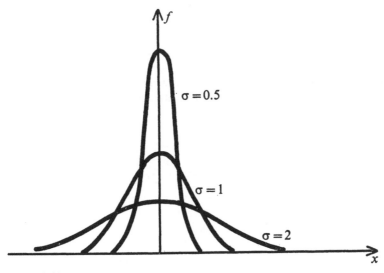

FIGURE 3.13

$$P(X = 0) = 0, \quad P(X \leq 0) = N(0; 0, 1), \quad \text{and} \quad P(X \leq 1) = N(1; 0, 1),$$

$$P(0 \leq X \leq 1) = P(X \leq 1) - P(X < 0) = P(X \leq 1) - P(X \leq 0) + P(X = 0)$$

$$= N(1; 0, 1) - N(0; 0, 1) = 0.8414 - 0.5000 = 0.3414.$$

To find $P(|X| \geq 1)$ we use (29) with $\mu = 0$:

$$P(|X| \geq 1) = 2[1 - N(1; 0, 1)] = 2(1 - 0.8414) = 0.3172.$$

EXERCISES

1 (a) Suppose we are given an urn containing two red balls and three white balls and suppose a sample of two balls is drawn at random without replacement. If $R(s)$ and $W(s)$ are respectively the number of red balls and the number of white balls in the sample $s$, then find

$$\{s \in S \mid R(s) \leq r\} = (R \leq r) \quad \text{and} \quad \{s \in S \mid W(s) \leq w\} = (W \leq w)$$

for all values of $r$ and $w$.

(b) Find $F_R(r)$ for all $r$ in this case.

2 Solve exercise 1 when the sample of two balls is drawn with replacement.

3 Show that $P(R = r) = b(r; 2, 2/5)$ in exercise 2.

4 Illustrate expressions (17) and (18) by drawing a positive-valued continuous function and depicting the various areas 'under the curve.'

5 Using a table for the normal distribution

$$(31) \quad N(x; 0, 1) = \int_{-\infty}^{x} \frac{e^{-t^2/2}}{\sqrt{2\pi}} \, dt,$$

with mean zero and variance 1, express each of the following as an algebraic expression involving instances of $N(x; 0, 1)$ for different values of $x$, and evaluate them:

(a) $\int_{-1}^{0} \frac{e^{-t^2/2}}{\sqrt{2\pi}} \, dt,$

(b) $\int_{1}^{+\infty} \frac{e^{-t^2/2}}{\sqrt{2\pi}} \, dt,$

(c) $\int_{-1}^{1} \frac{e^{-t^2/2}}{\sqrt{2\pi}} \, dt,$

(d) $\frac{1}{\sqrt{2\pi}} \left( \int_{-\infty}^{-1} e^{-t/2} \, dt + \int_{1}^{+\infty} e^{-t^2/2} \, dt \right).$

6 Consider a sequence of Bernoulli trials with probability of success in each trial equal to $p$. (a) Discuss how such a sequence of trials can be represented by random sampling with replacement from an urn containing $pN$ red balls and $N-pN$ white balls. (b) Let $X$ be the trial in which the first success occurs. Find $P(X = 0)$, $P(X = 1)$, $P(X = 2)$, and in general $P(X = n)$.

7 In exercise 3 of section 2.4, we investigated the probabilities involved in a dice game. This problem can be viewed in terms of random variables as follows. Let the probability space be as indicated in exercise 3 of section 2.4. Then let $Z$ be the function $Z(x, y) = x+y$.
    (a) Show that $Z$ is a random variable on $E$, $P$.
    (b) Find $P(Z = z)$ for all values that $Z$ can assume.
    (c) Find $F_Z(z)$ for all real $z$, and draw a graph of $F_Z(z)$.
    (d) If $X(x, y) = x$ and $Y(x, y) = y$, then show that $X$ and $Y$ are random variables.
    (e) Find $P(X = v)$ and $P(Y = v)$ for all values $v$ that $X$ and $Y$ can assume.
    (f) Find $F_X$ and $F_Y$.
    (g) Show that $F_Z(z) = \sum_{x+y \leq z} P(X = x) P(Y = y).$

8[1] In general if $X$ is a random variable with the property that $F_X(\lambda)$ is a differentiable function of $\lambda$, then $X$ possesses a *density function* $f_X(\lambda) = dF_X(\lambda)/d\lambda$ where

$$(32) \quad F_X(\lambda) = \int_{-\infty}^{\lambda} f_X(x) \, dx.$$

We have already encountered two examples of density functions in expressions (7) and (25). If $F_X(\lambda)$ is the CDF of a random variable, then assuming that $f_X(x)$ in (32) is a

---

1  The reader without calculus should ignore this problem.

continuous function show that it must have the following properties: (a) $f_X(x) \geq 0$ for all real $x$, (b) $\lim_{x \to +\infty} f_X(x) = 0$, (c) $\lim_{x \to -\infty} f_X(x) = 0$. Show that the density functions in (7) and (25) have these properties.

## 3.2
### THE MOMENTS OF A DISTRIBUTION

In section 3.1 we encountered various examples of random variables and their distributions. These examples fall naturally into four distinct classes:

(i) The random variable $V$ taking on a finite set of values, say $\{v_1, v_2, ..., v_m\}$, with a cumulative distribution function characterized by the $m$ numbers $P(V = v_i)$ for $i = 1, 2, ..., m$. (Example: the urn model at the beginning of section 3.1.)

(ii) The random variable $V$ taking on a countable set of values $v_1, v_2, ...$ with a CDF characterized by the sequence $\{P(V = v_i)\}_{i=1}^{\infty}$. (Example: the Poisson random variable to be discussed in section 3.3.)

(iii) The random variable $V$ taking on a continuous range of values $a \leq v \leq b$ with a density of probability equal to some non-negative function $f(v)$ different from zero for $v \in [a, b]$ so that

$$P(V \leq x) = F_V(x) = \int_{-\infty}^{x} f(v) \, dv = \int_{a}^{x} f(v) \, dv.$$

(Example: the uniform distribution of example 1 of section 3.1.)

(iv) The random variable $V$ taking on an infinite range of real values (usually either $(-\infty, \infty) = \{v \mid v \text{ is real}\}$ or $[0, \infty) = \{v \mid v \text{ is real and non-negative}\}$) with a density of probability given by some non-negative function $f(v)$ where

$$P(V \leq x) = F_V(x) = \int_{-\infty}^{x} f(v) \, dv.$$

(Example: the normal distribution discussed in example 4 of section 3.1.)

Technically there are other possibilities but these four are the ones usually encountered in practical situations. In each of these cases, and the others as well for that matter, we find that the CDF $F_V(x)$ must satisfy the following conditions:

(1) $0 \leq F_V(x) \leq 1$ for all $-\infty < x < \infty$,

(2) if $x_1 < x_2$, then $F_V(x_1) \leq F_V(x_2)$,

i.e., $F_V(x)$ is a *monotone non-decreasing function* of $x$,

(3) $\lim_{x \to -\infty} F_V(x) = 0$,

and

(4) $\lim_{x \to \infty} F_V(x) = 1$,

These properties follow directly from the axioms of a probability space given in chapter 2 and the definition of the cumultative distribution function.

If the random variable in question, $X$, is continuously distributed, it is often possible (indeed, in all cases we consider here) to determine a *density function* $f(v)$ for $X$ such that

$$F_X(v) = \int_{-\infty}^{x} f(v)\, dv.$$

From properties (1)–(4) above, we can infer that the following properties must be satisfied by such a density function $f$:

(5)  $0 \leqq f(v)$   for all $v \in (-\infty, \infty)$,

and

(6)  $\lim_{v \to -\infty} f(v) = \lim_{v \to +\infty} f(v) = 0.$

It is well to recall that classes (iii) and (iv) above can be treated together by assuming in (iii) that the density function $f(x)$ is defined for all real $v$ but is non-zero only when $a \leqq v \leqq b$. Thus

$$\int_{-\infty}^{x} f(v)\, dx = \begin{cases} 0 & \text{for } x < a, \\ \int_{a}^{x} f(v)\, dx & \text{for } a \leqq x \leqq b, \\ 1 & \text{for } b < x. \end{cases}$$

### 3.2.1 *The mean and variance*

Using the analogy of the distribution of mass as we did with a sample in chapter 1, we can construct two measures, one an indication of the 'centre' of the distribution and the other a measure of the 'spread' of the distribution, which in some cases determine the distribution completely.

Recall that the centre of gravity of a mass distribution, say on a line, is the point where the moment (or torque) engendered by mass on the left exactly balances the moment engendered by the mass on the right of the point. Thus if a knife-edge were placed at that point, the mass distribution would be exactly in balance.

For the continuous probability (mass) distribution, this point, called the *mean* of the distribution, is given by the equation

(7)  $E(X) = \mu = \int_{-\infty}^{\infty} vf(v)\, dv$

where $f(v)$ is the density function of the distribution. This value is sometimes called the *expected value* of $X$ and is denoted either by $E(X)$ or by $\mu$.

If $X$ takes on only a finite set of values, then the expected value of $X$ and the mean of its distribution are both given by

(8) $E(X) = \mu = \sum_{i=1}^{n} a_i P(X = a_i)$

where $a_1, a_2, ..., a_n$ are all the different values that $X$ can assume.

If $X$ has a discrete distribution but takes on an infinite set of values $a_1, a_2, ..., a_n, ...,$ then[2]

(8a) $E(X) = \mu = \sum_{i=1}^{\infty} a_i P(X = a_i) \overset{d}{=} \lim_{n \to \infty} \sum_{i=1}^{n} a_i P(X = a_i)$

provided that the series in (8a) converges, i.e., that the limit exists as $n \to \infty$.

Thus, if $X$ is normally distributed with

$$F_X(v) = N(v; \mu, \sigma),$$

then by (7) we have

$$E(X) = \frac{1}{\sigma\sqrt{2\pi}} \int_{-\infty}^{\infty} x \, e^{-(x-\mu)^2/2\sigma^2} \, dx,$$

and using a short argument involving the calculus, we obtain the result

$$E(X) = \mu.$$

(In section 3.1, $\mu$ was called the mean of the normal distribution $N(x; \mu, \sigma)$ in anticipation of the definition of the distribution mean given here.)

If $X$ is a binomially distributed random variable as introduced in example 3 of section 3.1 so that

$$P(X = k) = \binom{n}{k} p^k (1-p)^{n-k}$$

for $k = 0, 1, 2, ..., n$, then

$$E(X) = \sum_{k=0}^{n} k \, \frac{n!}{k!\,(n-k)!} p^k (1-p)^{n-k}$$

or, after a rearrangement of the factor involving factorials,

(9) $E(X) = \sum_{k=1}^{n} np \, \frac{(n-1)!}{(k-1)!\,(n-1-(k-1))!} p^{k-1} (1-p)^{n-1-(k-1)}.$

Let $i = k-1$; then (9) can be rewritten

$$E(X) = \sum_{i=0}^{n-1} np \, \frac{(n-1)!}{i!\,(n-1-i)!} p^i (1-p)^{n-1-i}$$

$$= np \sum_{i=0}^{n-1} b(i; n-1, p) = np \cdot 1 = np.$$

2  The symbol $\overset{d}{=}$ stands for 'is defined equal to.'

Thus the mean of a binomial random variable $X$ is $\mu = np$.

Just as in section 1.6 where we defined $\mathcal{M}(g(X))$, the arithmetic mean (for a sample) of a function $g$ of $X$, we can define

(10) $E(g(X)) = \int_{-\infty}^{\infty} g(v)f(v)dv$

if $X$ possesses a density function $f(v)$ or

(11) $E(g(X)) = \sum_{i=1}^{Q} g(a_i)P(X = a_i)$

if $X$ takes on only the values $a_i$ for integers $1 \leq i \leq Q$, where $Q$ equals $n$ if $X$ assumes only $n$ values or $Q$ equals $\infty$ if $X$ can assume an infinite set of values $a_i$ for $i = 1, 2, \ldots$, whichever is appropriate.

The distributional analogue of the sample variance, i.e., the distributional variance, is given by (10) or (11), whichever is appropriate, when $g(x) = (x - E(X))^2$. As was the case for the sample variance, the *distributional variance* is a measure of the spread of the distribution and is given by

(10′) $\sigma_X^2 = \int_{-\infty}^{\infty} (x-\mu)^2 f(x)\, dx$

for a continuously distributed random variable $X$ with density function $f(x)$, and by

(11′) $\sigma_X^2 = \sum_{i=1}^{Q} (a_i-\mu)^2 P(X = a_i)$

for a discretely distributed random variable which takes on the values $a_i$ only.

In either case,

$$\sigma_X^2 = E[(X-E(X))^2] = E(X^2-2E(X)X+E(X)^2)$$
$$= E(X^2)-2E(X)E(X)+[E(X)]^2,$$

and so

(12) $\sigma_X^2 = E(X^2)-[E(X)]^2$.

Expression (12) is useful for computing the distributional variance in certain special cases.

For a random variable $X$ with the normal distribution $N(x; \mu, \sigma)$, it can be shown using the calculus that

$$\sigma_X^2 = \frac{1}{\sigma\sqrt{2\pi}} \int_{-\infty}^{\infty} (x-\mu)^2 e^{-(x-\mu)^2/2\sigma^2}\, dx$$

is equal to $\sigma^2$.

Similarly, if $X$ is binomially distributed, then by definition

$$\sigma_X{}^2 = \sum_{i=1}^{n} (i-np)^2 \binom{n}{i} p^i (1-p)^{n-i}.$$

After some calculation, it turns out that

$$\sigma_X{}^2 = np(1-p)$$

in this case.

Note that in these two cases the distribution is completely determined by the mean and variance. For the normal distribution these appear explicitly in the formula, while for the binomial we have

$$\mu = np, \qquad \sigma^2 = np(1-p),$$

and so

$$p = \frac{\mu - \sigma^2}{\mu}, \qquad n = \frac{\mu^2}{\mu - \sigma^2}$$

There are some distributions where the mean and variance do not exist. In the *Cauchy distribution* neither exists. A random variable $X$ is said to have the Cauchy distribution if

$$(13) \quad F_X(x) = \frac{1}{\pi} \int_{-\infty}^{x} \frac{1}{1+v^2} \, dv.$$

The expression for the mean of $F_X(x)$ can be shown to take the form

$$\mu = E(X) = \frac{1}{\pi} \int_{-\infty}^{\infty} \frac{x}{1+x^2} \, dx = \frac{1}{2\pi} \lim_{\substack{a \to \infty \\ b \to \infty}} [\log(a^2) - \log(b^2)]$$

where the limit on the right does not exist. Since we do not have a mean, the variance cannot be computed either.

Strangely enough the graph of the Cauchy distribution density function is symmetric about the line $X = 0$ and looks very much like the graph of the density function of the unit normal distribution $N(x; 0, 1)$:

$$f(x) = \frac{1}{\sqrt{2\pi}} e^{-x^2/2}$$

### 3.2.2 *Higher moments*

Just as in section 1.6, where the sample mean and sample variance were special cases of sample moments about the origin and central sample moments, so the distributional mean and variance are special cases of distribution moments about the origin and the mean.

In general $E(x^r)$ is called the *rth moment* of (the distribution of) the random variable $X$ *about the origin*, and $\mu_r(X) = E[(X - E(X))^r]$ is the *rth central moment* of $X$, sometimes called the *rth moment* of the distribution of $X$ *about the mean of X*.

We shall see in chapter 4 how the *rth* sample moments can be used to approximate the *rth* distributional moments.

As in section 1.6 where we obtained measures of the skewness and peakedness for the sample histogram using the sample moments, we can use the moments of the distribution of $X$ to obtain analogous measures for the distribution. Thus

$$\beta_1(X) = \frac{[E\{(X - E(X))^3\}]^2}{[\sigma_X{}^2]^3}$$

is a measure of the skewness of the distribution of $X$. Symmetry obtains if $\beta_1(X) = 0$; if $\mu_3(X) < 0$, the distribution is skewed to the left; and if $\mu_3(X) > 0$, it is skewed to the right.

In an analogous fashion

$$\beta_2(X) = \frac{E\{(X - E(X))^4\}}{[\sigma_X{}^2]^2}$$

is a measure of kurtosis.

It can be shown that if $X$ is normally distributed with mean $\mu$ and variance $\sigma$, then

$$\beta_1(X) = 0, \qquad \beta_2(X) = 3.$$

If $X$ has a binomial distribution,

$$\beta_1(X) = \frac{(1 - 2p)^2}{np(1 - p)}, \qquad \beta_3(X) = 3 + \frac{1 - 6p(1 - p)}{np(1 - p)}.$$

## EXERCISES

1 Show that each of the CDF's defined by (a) expression (3) of section 3.1, (b) expressions (5) and (6) of section 3.1, (c) expression (21) of section 3.1, satisfies expressions (1)–(4) in this section.

2[3] Show that each of the CDF's of (a) the Cauchy distribution and (b) the normal distribution satisfies expressions (1)–(4).

3 Find $E(X)$ and $\sigma_X{}^2$ when $X$ is uniformly distributed on the interval $[a, b]$.

4 If $X$ has the binomial distribution with parameters $n$ and $p$, find $E(X^2)$ and $E\{(X - E(X))^2\}$.

---

3   The reader without calculus should ignore this problem.

5 Under what circumstances does $\beta_1(X) = 0$ when $X$ has the binomial distribution?

6 If we use the notation $\mu_r = E[(X - E(X))^r]$ and $m_r = E(X^r)$, show that

$$\mu_r = \sum_{t=0}^{r} (-1)^t \binom{r}{i} m_{r-t} m_1^t$$

for $r = 2, 3, \dots$

7 If $X$ has the hypergeometric distribution so that

$$P(X = x) = \frac{\binom{M}{x} \binom{N-M}{n-x}}{\binom{N}{n}} = h(x; n, M, N),$$

then $E(X) = nM/N$ and

$$\sigma_x^2 = \frac{nM(N-M)(N-n)}{N^2(N-1)}.$$

Show that if $N$ gets large with $M/N = p$ fixed, then for small $n$

$$\lim_{N \to \infty} E(X) = np \quad \text{and} \quad \lim_{N \to \infty} \sigma_x^2 = np(1-p).$$

Explain the phenomenon in terms of an urn model.

## 3.3
### SOME SPECIAL DISTRIBUTIONS

To this point we have considered only a few of the particular distributions useful in language and style studies: the binomial and hypergeometric (discrete) distributions, and the uniform, normal, and Cauchy (continuous) distributions.

Among the discrete distributions one of primary importance is the *Poisson distribution*. A random variable $X$ has the Poisson distribution with parameter $\lambda > 0$ if

(1) $\quad P(X = x) = p(x; \lambda) = \dfrac{\lambda^x}{x!} e^{-\lambda}$

for $x = 0, 1, 2, 3, \dots$ This is an example of category (ii) mentioned at the beginning of section 3.2, for $x$ can take on any non-negative integer value.

Random variables whose values depend on rare chance events follow the Poisson distribution: for example, (i) the number of misprints per page in a large volume of (comparable) printed material, (ii) the number of accidents per day in a large city, (iii) the number of calls per minute at a telephone switchboard. Of more interest to us, the number of instances of articles (*a, an, the*) per

TABLE 3.2

| Term | Contribution to $e^{-0.3}$ | TOTAL |
|------|------|------|
| 1 | 1.0000 | 1.0000 |
| 2 | −0.3000 | 0.7000 |
| 3 | 0.0450 | 0.7450 |
| 4 | −0.0045 | 0.7405 |
| 5 | 0.0003 | 0.7408 |
| 6 | −0.00002 | 0.74078 |

$N$-word passage of text (for $50 \leq N \leq 100$), among certain authors at least, gives strong evidence of having the Poisson distribution. We shall return to this point later.

If $X$ has the Poisson distribution with parameter $\lambda$, then its CDF is $F_X(x)$ where $F_X(x) = 0$ for $x < 0$ and

$$(2) \quad F_X(x) = \sum_{k=0}^{[x]} \frac{\lambda^k e^{-\lambda}}{k!}$$

for $x \geq 0$. In this expression, $[x]$ stands for the largest integer not greater than $x$, sometimes called the *integral part of x*. In order to check whether $F_X(x)$ satisfies the properties (1)–(4) in section 3.2 that are required of a CDF, we need to discover some of the properties of the *exponential function* $e^x$. Tables of the exponential function $e^x$ are easy to find, so that (1) and (2) are easily calculated. It might be well to note, however, that $e^x$ is also easily calculated from its series expansion[4]

$$(3) \quad e^x \overset{d}{=} \sum_{k=0}^{\infty} \frac{x^k}{k!} = 1 + x + \frac{x^2}{2!} + \frac{x^3}{3!} + \frac{x^4}{4!} + \cdots$$

If, for example, we wish $e^{-0.3}$ to, say, four decimal places, we merely copy out the appropriate terms of (3) until the first five decimal places become stable, as in table 3.2. It is clear at this point that the seventh and succeeding terms will not affect the first five decimal places and so

$$e^{-0.3} = 0.7408$$

when rounded off to the first four places.

To compute from (1), with $\lambda = 0.3$, the probability that $X = 2$, we form

$$(4) \quad P(X = 2) = (0.3)^2(0.7408)/2! = (0.09)\,(0.7408)/2 = 0.0333.$$

To check that the Poisson variate $X$ has a CDF $F_X(x)$ with the desired properties, note that conditions (1), (2), and (3) in section 3.2 are easily verified for such an $F_X(x)$. Condition (4) is all that might be in doubt. However

The symbol $\overset{d}{=}$ stands for 'is defined equal to.'

$$\lim_{x \to \infty} F_X(x) = \lim_{m \to \infty} \sum_{k=0}^{m} \frac{\lambda^k e^{-\lambda}}{k!} = e^{-\lambda} \sum_{k=0}^{\infty} \frac{\lambda^k}{k!}$$

$$= e^{-\lambda} e^{\lambda} = e^{-\lambda + \lambda} = e^0 = 1.$$

Therefore (4) also holds and $F_X(x)$ as defined in expression (2) is indeed a CDF.

The Poisson distribution appears as a limiting distribution of the binomial distribution, as the following example illustrates.

EXAMPLE 1  In example 3 of section 3.1, we treated the random variable $Z$, the number of occurrences of /ž/ in a 1000-phoneme corpus, and found that

(5)  $P(Z = k) = \binom{1000}{k} (0.0003)^k (0.9997)^{1000-k}.$

Since the probability of /ž/ is very low and the number of trials is very high in (5), the computation of this expression is somewhat arduous. However, it has been shown that if $np = \lambda$ *remains a fixed constant as* $n \to \infty$, *then*

$\lim_{n \to \infty} b(k; n, p) = \lambda^k e^{-\lambda}/k!$

*for* $k = 0, 1, 2, \ldots$ Thus[5] for large values of $n$ and small values of $p$,

$b(k; n, p) \approx \lambda^k e^{-\lambda}/k!$

where $\lambda = np$, and so (5) can be expressed approximately as

(6)  $P(Z = k) = (0.3)^k e^{-0.3}/k!$

Using expression (4), we find for example that

$P(Z = 2) = 0.0333.$

To check the degree of approximation obtained we recalculate $P(Z = 0)$ using the Poisson approximation

$P(Z = 0) = (0.3)^0 e^{-0.3}/0! = e^{-0.3} = 0.7408.$

The result obtained in example 3 of section 3.1 was 0.7413; hence, carried out to three decimal places, the Poisson approximation gives the actual value.

If $X$ has the Poisson distribution with parameter $\lambda$, then we can show that $\lambda$ is in fact the distributional mean as follows:

$$E(X) = \sum_{k=0}^{\infty} kP(X = k) = \sum_{k=0}^{\infty} \frac{k\lambda^k e^{-\lambda}}{k!} = \sum_{k=1}^{\infty} \frac{\lambda^k e^{-\lambda}}{(k-1)!}$$

$$= \sum_{j=0}^{\infty} \frac{\lambda^{j+1} e^{-\lambda}}{j!} = \lambda \sum_{j=0}^{\infty} \frac{\lambda^j e^{-\lambda}}{j!} \quad (\text{with } j = k-1)$$

$$= \lambda \lim_{x \to \infty} F_X(x) = \lambda.$$

5  The symbol $\approx$ signifies 'is approximately equal to.'

Similarly, we can show that

$$\sigma_X{}^2 = \lambda, \quad \beta_1(X) = 1/\lambda, \quad \beta_2(X) = 3 + 1/\lambda.$$

Another discrete distribution that is useful in linguistic work (see Mosteller and Wallace 1964) is the *negative binomial distribution*:

$$(7) \quad P(X = x) = n(x; k, p) = \frac{k(k+1)\ldots(k+x-1)}{x!} p^k(1-p)^x$$

where $k > 0$ and $0 \leq p \leq 1$ are parameters, and $x = 0, 1, 2, \ldots$

Under certain circumstances the Poisson distribution $p(x; \lambda)$ is a good approximation to the distribution of the number $X$ of occurrences of a word $W$ in a block of text of length $N$ when the average number of occurrences over the whole text is $\lambda$. However on some occasions the negative binomial distribution provides a better approximation. See Mosteller and Wallace (1964, section 2.3) for a discussion of this point.

Perhaps the reader is wondering how the appropriate values of $\lambda$, in the case of the Poisson distribution, and $k$ and $p$, in the case of the negative binomial distribution, are to be chosen. This is a problem of the *estimation of parameters*, which is the subject of the next chapter.

One of the methods for estimating parameters involves a knowledge of various moments of the distribution in question. For a random variable $X$ with the negative binomial distribution $n(x; k, p)$, we have

$$E(X) = \frac{k(1-p)}{p},$$

$$\sigma_X{}^2 = \frac{k(1-p)}{p^2},$$

$$\beta_1(X) = \frac{p(2-p)}{k(1-p)},$$

and

$$\beta_2(X) = 3 + \frac{p^2}{k(1-p)} + \frac{6}{p}.$$

When $k = 1$, the negative binomial distribution reduces to the *geometric* or *Pascal* distribution:

$$(8) \quad P(X = x) = g(x; p) = p(1-p)^x$$

for $x = 0, 1, 2, \ldots$ If a sequence of Bernoulli trials are made where the probability of an event $E$ in a particular trial is $p$ and $X$ is the number of the trial upon which the first success occurs, then the distribution of $X$ is given by (8). See exercise 6 of section 3.1.

TABLE 3.3

| No. of personal pronouns per passage | No. of passages with that no. of personal pronouns |
|:---:|:---:|
| 0 | 1 |
| 1 | 5 |
| 2 | 2 |
| 3 | 7 |
| 4 | 10 |
| 5 | 3 |
| 6 | 5 |
| 7 | 6 |
| 8 | 4 |
| 9 | 4 |
| 10 | 1 |
| 11 | 2 |

EXAMPLE 2   In a sample of fifty 50-word passages selected at random from *Riders in the Chariot* by Patrick White, we obtained the data in table 3.3. Here the personal pronouns are taken to be *I, me, my, mine, you, your, yours, he, him, his, she, her, hers, it, its, we, us, our, ours, they, them, their, theirs*. The sample mean and variance are $\bar{X} = 5.140$ and $S_X^2 = 7.878$ where the random variable $X$ is taken to be the number of personal pronouns per passage. There is good evidence that $X$ has the negative binomial distribution with $p = 0.652$ and $k = 9.649$.

The computation of the negative binomial probabilities offers some difficulty to those without access to a computer because, in this case for example, it involves the calculation of

$$p^k = (0.652)^{9.649}.$$

The easiest way to obtain this value is to use logarithms (see section 3.1). To find $p^k$ first form $\log_{10} p$, which can be obtained from a table of logarithms. Then

$$\log_{10} p^k = k \log_{10} p.$$

In our case

$$\log_{10} (0.652)^{9.649} = 9.649 \log 0.652 = 9.649(9.814248 - 10)$$

$$= 7.20768 - 9.$$

The number with this logarithm is 0.0161. Thus

$$P(X = 0) = p^k = 0.0161,$$

$$P(X = 1) = p^k k(1-p) = 0.0542,$$

$$P(X = 2) = p^k [k(k+1)/2] (1-p)^2 = 0.1004,$$

and so

$$F_X(2) = \sum_{k=0}^{2} P(X = k) = 0.1707.$$

Thus the expected number of sample-passages which yield a value of $X$ less than or equal to 2 is $P(X \leq 2)$ times 50 or

$$F_X(2)\,(50) = (0.1707)\,(50) = 8.54.$$

From table 3.3, the actual value obtained in this sample is 8. The reader is asked in exercise 6 below to find the remaining expected numbers.

A class of continuous distributions of some interest is the *gamma distributions*. A random variable $X$ is said to have a gamma distribution with parameters $\alpha$ and $\beta$ if the CDF of $X$ takes the form

$$(9)\quad F_X(x) = \begin{cases} 0 & \text{if } x < 0, \\ \displaystyle\int_0^x \frac{t^\alpha e^{-t/\beta}}{\beta^{\alpha+1}\Gamma(\alpha+1)}\,dt & \text{otherwise} \end{cases}$$

$$= \gamma(x;\,\alpha,\,\beta)$$

where $\alpha > -1$ and $\beta > 0$. The symbol $\Gamma(x)$ stands for the *gamma function* of $x$,

$$(10)\quad \Gamma(x) = \int_0^\infty t^{x-1} e^{-t}\,dt$$

defined for $x > 0$. If $x$ is a positive integer, then it can be shown that

$$(11)\quad \Gamma(x) = (x-1)!$$

and in general

$$(12)\quad \Gamma(x) = (x-1)\Gamma(x-1).$$

The gamma function and the gamma distribution with $\beta = 1$ are extensively tabulated (Pearson and Hartley 1962). Figure 3.14 shows some examples of density functions of the gamma distribution for various values of $\alpha$ and $\beta$.

Because it is possible to show that

$$(13)\quad \gamma(x;\,\alpha,\,\beta) = \gamma(x/\beta;\,\alpha,\,1),$$

these tables can also be used to evaluate $F_X(x)$ when $\beta \neq 1$.

The special case when $\alpha = 0$ is of some interest to us. Then

$$(14)\quad F_X(x) = \gamma(x;\,0,\,\beta) = \begin{cases} \dfrac{1}{\beta}\displaystyle\int_0^x e^{-t/\beta}\,dt = 1 - e^{-x/\beta} & \text{for } x \geq 0, \\ 0 & \text{for } x < 0. \end{cases}$$

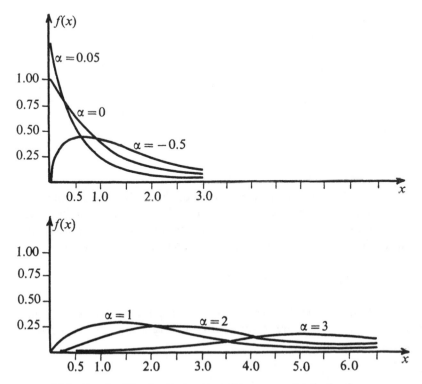

FIGURE 3.14 Graphs of the density function of the gamma distribution for various values of $\alpha$ and $\beta = 1$.

A random variable with this CDF is said to have the *exponential distribution* with parameter $\beta$. A simple derivation using calculus yields the results:

$$E(X) = \beta, \quad \sigma_X{}^2 = \beta^2, \quad \beta_1(X) = 4, \quad \beta_2(X) = 9.$$

The length of life (i.e., hours of service before burning out) of light bulbs and that of vacuum tubes have the exponential distribution.

Patterns of pauses and bursts of conversational speech are also describable in terms of the exponential distribution. For example, if $X$ is the length of a silence in an ordinary conversation, then over a large sample we find that

$$P(X \leq x) = F_X(x) = 1 - e^{-x/\beta}$$

to a good approximation for a suitable choice of $\beta$ (Cassotta et al. 1964).

In example 1 we have seen how, under certain circumstances, as $n \to \infty$ the Poisson distribution can be used to approximate the more difficult to calculate binomial distribution $b(x; n, p)$. The normal distribution can also be used to approximate the binomial distribution. Indeed, we shall show in corollary 1 of section 3.5 that *for fixed p*

E

(15) $\lim_{n \to \infty} b(x; n, p) = \dfrac{1}{\sqrt{npq}\ \sqrt{2\pi}}\ e^{-(x-np)^2/2npq}$

where $q = 1 - p$. In general, this approximation is good if $np \geqq 5$.

We cannot use (15) to approximate $P(X = 2)$ in example 1 because $np = 0.3$ is too small. However, the following example provides a case which is capable of normal approximation and indicates how we compensate for the fact that the binomial random variable is discrete while a normal random variable is continuous.

EXAMPLE 3   Roberts found (see our table 2.4) that the relative frequency of /ɔ/ in his sample was 0.0064. If we assume this value to be the probability of obtaining /ɔ/ when a phoneme is selected at random from a text, then, assuming independence in addition, we find that the probability that the number of occurrences $X$ of /ɔ/ in a 10,000-phoneme corpus equals $k$ is given by the formula

(16) $P(X = k) = \dbinom{10{,}000}{k} (0.0064)^k (0.9936)^{10{,}000-k}.$

Since $np = 10{,}000 \times 0.0064 = 64 > 5$, we can use the normal approximation. The conventional procedure when approximating a discrete variate by a continuous one is to compute

(17) $P(k - \tfrac{1}{2} \leqq X \leqq k + \tfrac{1}{2})$

using the approximating distribution. Under the assumption that $X$ is normally distributed with mean $np = 64$ and variance $npq = 63.59$, we obtain

$$
\begin{aligned}
\text{(18) } P(k - \tfrac{1}{2} \leqq X \leqq k + \tfrac{1}{2}) &= P(X \leqq k + \tfrac{1}{2}) - P(X < k - \tfrac{1}{2}) \\
&= P(X \leqq k + \tfrac{1}{2}) - P(X \leqq k - \tfrac{1}{2}) \\
&= N(k + \tfrac{1}{2}; 64, 63.59) - N(k - \tfrac{1}{2}; 64, 63.59)
\end{aligned}
$$

because $P(X = k - \tfrac{1}{2}) = 0$, which follows because $X$ is assumed to be normally and hence continuously distributed (see exercise 5). We have already observed in section 3.1 that the area

(19) $N(x; 0, 1) = \dfrac{1}{\sqrt{2\pi}} \displaystyle\int_{-\infty}^{x} e^{-t^2/2}\, dt$

under the curve

$$ f(x) = \dfrac{1}{\sqrt{2\pi}}\ e^{-t^2/2} $$

between $-\infty$ and $x$ can be obtained from tables (e.g., Pearson and Hartley, 1962, p. 1 and table 1). In addition, a technical computation shows that

(20) $N(x; \mu, \sigma) = N\left(\dfrac{x-\mu}{\sigma}; 0, 1\right).$

Therefore from (19) we obtain

(21) $P(k-\tfrac{1}{2} \leq X \leq k+\tfrac{1}{2}) = N\left(\dfrac{k-63.5}{\sqrt{63.59}}; 0, 1\right) - N\left(\dfrac{k-64.5}{\sqrt{63.59}}; 0, 1\right).$

Suppose we want to find $P(X = 60)$. Using the normal approximation and expression (21), we have

$$
\begin{aligned}
P(X = 60) &= N\left(\dfrac{60-63.5}{\sqrt{63.59}}; 0, 1\right) - N\left(\dfrac{60-64.5}{\sqrt{63.59}}; 0, 1\right) \\
&= N\left(-\dfrac{3.5}{7.97}; 0, 1\right) - N\left(-\dfrac{4.5}{7.97}; 0, 1\right) \\
&= N(-0.44; 0, 1) - N(-0.56; 0, 1) \\
&= [1 - N(0.44; 0, 1)] - [1 - N(0.56; 0, 1)] \\
&= N(0.56; 0, 1) - N(0.44; 0, 1) \\
&= 0.7123 - 0.6736 = 0.039 \text{ (approximately)}.
\end{aligned}
$$

The normal distribution turns out to be a limiting distribution in the above sense for many other distributions. In addition, we shall see in section 3.5 that the sum of $n$ observations taken at random from almost any population tends to become normal as $n$ increases.

EXERCISES

1 Using expression (6) find (a) $P(Z = 0)$, (b) $P(Z = 1)$, (c) $P(Z = 3)$, (d) $P(Z = 10)$ to four decimal places.

2 In order to test the Poisson approximation, use expression (23) in section 3.1 to obtain $P(Z = 1)$, $P(Z = 3)$, and $P(Z = 4)$.

3 Derive the variance of $X$, $\beta_1(X)$, and $\beta_2(X)$ when $X$ has the Poisson distribution with mean $\lambda$.

4 A distribution related to the Poisson distribution is the *truncated Poisson distribution*:

$$
P(X = x) = \dfrac{\lambda^x e^{-\lambda}}{(1 - e^{-\lambda})x!} = \dfrac{\lambda^x}{(e^{\lambda} - 1)x!}
$$

ɪ $x = 1, 2, \ldots$

(a) Show that the CDF

$$F_X(x) = \begin{cases} 0 & \text{for } x < 1, \\ \displaystyle\sum_{k=1}^{[x]} \frac{\lambda^x}{(e^\lambda - 1)x!} & \text{for } x \geq 1 \end{cases}$$

of $X$ has properties (1)–(4) in section 3.2.

(b) Show that $E(X) = \lambda/(1 - e^{-\lambda})$.

5 Knowing that for $|x| < 1$

$$\sum_{k=0}^{\infty} x^k = \frac{1}{1-x},$$

$$\sum_{k=0}^{\infty} kx^{k-1} = \frac{1}{(1-x)^2},$$

...

$$\sum_{k=0}^{\infty} \frac{k(k-1)...(k-m+1)}{m!} x^{k-m} = \frac{1}{(1-x)^{m+1}},$$

find (a) $E(X)$, (b) $E(X^2)$, (c) $E(X^3)$ for a random variable $X$ with geometric distribution $g(x; p)$ as defined in expression (8).

6 Calculate the expected number, $50P(X = x)$, of passages with $x$ personal pronouns for $x = 0, 1, 2, ..., 11$ in example 2 and compare your results with the observed frequencies. In chapter 5 we shall obtain a method for testing the 'goodness of the fit' to the observed data achieved by these theoretical frequencies.

## 3.4
### JOINT DISTRIBUTIONS AND THE STANDARD ERROR OF PROPORTION

#### 3.4.1 *Joint distributions*

To this point, we have considered only one random variable at a time. Sometimes, however, there is a great advantage in being able to consider more than one and to consider their interaction.

Let $X_1, X_2, ..., X_n$ be $n$ random variables. Suppose they are all defined on the same probability space so that the event

$$(X_1 \leq \lambda_1, X_2 \leq \lambda_2, ..., \text{and } X_n \leq \lambda_n) = \bigcap_{k=1}^{n} (X_k \leq \lambda_k)$$

has a probability for any choice of $\lambda_1, \lambda_2, ..., \lambda_n$. Then we define the *joint cumulative* distribution function of the $n$-tuple $(X_1, X_2, ..., X_n)$ of random variables to be the function

(1) $\quad F_{X_1, X_2, \ldots, X_n}(\lambda_1, \lambda_2, \ldots, \lambda_n) = P(X_1 \leq \lambda_1, X_2 \leq \lambda_2, \ldots, X_n \leq \lambda_n)$

$$= P\left(\bigcap_{k=1}^{n} (X_k \leq \lambda_k)\right)$$

for $-\infty < \lambda_i < \infty$ $(i = 1, 2, \ldots, n)$.

EXAMPLE 1   In the dice-throwing problem discussed in exercises 3 of section 2.4 and 7 of section 3.1, we distinguished the dice and considered the probability space $E, P$ where the set of outcomes was

$$U = \{(x, y) \mid x = 1, 2, \ldots, 6 \text{ and } y = 1, 2, \ldots, 6\}.$$

$E$ was the set of all subsets of $U$, and for any $A \in E$, we assigned the probability

$$P(A) = |A|/36.$$

Thus, for example, if $A = \{(1, 2)\}$, then the probability of obtaining 1 on the first die and 2 on the second is

$$P(A) = P\{(1, 2)\} = 1/36,$$

and the probability of $\{(1, 2), (2, 1)\}$ is

$$P\{(1, 2), (2, 1)\} = 2/36 = 1/18.$$

The functions $X(x, y) = x$ and $Y(x, y) = y$ are random variables and

$$P(X = x \text{ and } Y = y) = P(X = x, Y = y) = P\{(x, y)\} = 1/36.$$

Thus the joint cumulative distribution of $(X, Y)$ is given by the expression

$$F_{X,Y}(\lambda_1, \lambda_2) = P(X \leq \lambda_1, Y \leq \lambda_2) = \sum_{\substack{x \leq \lambda_1 \\ y \leq \lambda_2}} P\{(x, y)\}.$$

In particular, if $\lambda_1 = 1\frac{1}{2}$ and $\lambda_2 = 5\frac{1}{2}$,

$$F_{X,Y}(1\tfrac{1}{2}, 5\tfrac{1}{2}) = \sum_{\substack{x \leq 1\frac{1}{2} \\ y \leq 5\frac{1}{2}}} P\{(x, y)\}$$

$$= P\{(1, 1)\} + P\{(1, 2)\} + P\{(1, 3)\} + P\{(1, 4)\} + P\{(1, 5)\}$$

$$= 5/36.$$

Since for any $i, j$ the event

$$(X = i, Y = j) = \{(x, y) \mid X(x, y) = i \text{ and } Y(x, y) = j\}$$

$$= (x = i) \cap (Y = j) = \{(i, j)\},$$

while

$$P(X = i, Y = j) = P\{(i, j)\} = 1/36$$

and

$$P(X = i) = 6/36 \quad \text{and} \quad P(Y = j) = 6/36,$$

it follows that

$$P(X = i, Y = j) = P(X = i)P(Y = j),$$

and so the events $(X = i)$ and $(Y = j)$ are independent for any choice of $i$ and $j$. We can therefore write

$$F_{X,Y}(\lambda_1, \lambda_2) = \sum_{\substack{x \leq \lambda_1 \\ y \leq \lambda_2}} P(X = x, Y = y) = \sum_{\substack{x \leq \lambda_1 \\ y \leq \lambda_2}} P(X = x)P(Y = y)$$

$$= \left( \sum_{x \leq \lambda_1} P(X = x) \right) \left( \sum_{y \leq \lambda_2} P(Y = y) \right) = F_X(\lambda_1)F_Y(\lambda_2).$$

In general a pair of random variables $X$, $Y$ is said to be *independent* provided that for all $\lambda_1$ and $\lambda_2$

(2) $\quad F_{X,Y}(\lambda_1, \lambda_2) = F_X(\lambda_1)F_Y(\lambda_2).$

It is clear from example 1 that in that case $X$ and $Y$ are independent.

EXAMPLE 2 Let $X$ and $Y$ be two independent random variables with uniform distribution on $[0, 1]$. Then, as is illustrated in figure 3.15,

$$F_{X,Y}(x, y) = F_X(x)F_Y(y)$$

$$= \begin{cases} 0 & \text{if either } x \text{ or } y \text{ is } <0, \\ xy & \text{for } 0 \leq x \leq 1 \text{ and } 0 \leq y \leq 1, \\ x & \text{for } 0 \leq x \leq 1 \text{ and } y > 1, \\ y & \text{for } 0 \leq y \leq 1 \text{ and } x > 1, \\ 1 & \text{otherwise.} \end{cases}$$

Because (2) holds, we can express $F_{X,Y}(x, y)$ as a product of integrals

$$F_{X,Y}(x, y) = \left( \int_{-\infty}^{x} f_X(t)\, dt \right) \left( \int_{-\infty}^{y} f_Y(t)\, dt \right)$$

where

$$f_X(t) = f_Y(t) = \begin{cases} 0, & 0 < t, \\ 1, & 0 \leq t \leq 1, \\ 0, & 1 < t. \end{cases}$$

In what follows, we shall consider only the definitions for discrete random variables; the reader can construct the analogous definitions for continuously distributed random variables if he wishes.

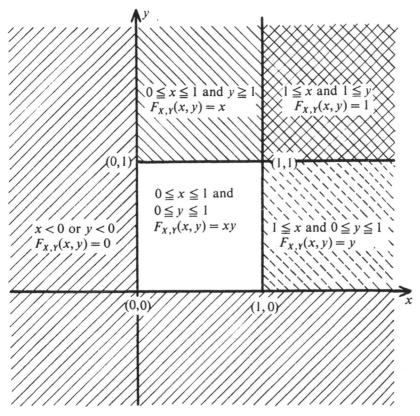

FIGURE 3.15   The various values of $F_{X,Y}(x, y)$ in example 2.

In general, if $X_1, X_2, \ldots, X_n$ is a set of random variables whose range in each case is the set of non-negative integers, then their joint cumulative distribution function is given by the expression

(3)   $F_{X_1, X_2, \ldots, X_n}(\lambda_1, \lambda_2, \ldots, \lambda_n) = \sum_{\substack{i_1 \leq \lambda_1 \\ i_2 \leq \lambda_2 \\ \cdots \\ i_n \leq \lambda_n}} P(X_1 = i_1, X_2 = i_2, \ldots, X_n = i_n),$

and for a real-valued function $g(x_1, x_2, \ldots, x_n)$ of $n$ real variables, the expectation of

$$g(X_1, X_2, \ldots, X_n)$$

is given by the expression

(4)   $E[g(X_1, X_2, \ldots, X_n)]$

$$= \sum_{\substack{i_1 \in \text{range of } X_1 \\ i_2 \in \text{range of } X_2 \\ \cdots \\ i_n \in \text{range of } X_n}} g(i_1, i_2, \ldots, i_n)p(X_1 = i_1, X_2 = i_2, \ldots, X_n = i_n).$$

As in example 1, we can show in general that if two random variables $X$ and $Y$ are independent, non-negative, and integer-valued, then

(5)  $P(X = i, Y = j) = P(X = i)P(Y = j)$.

From (4) and (5) it follows that

$$E(X+Y) = \sum_{\substack{i \in \text{ range of } X \\ j \in \text{ range of } Y}} (i+j)P(X=i)P(Y=j)$$

$$= \sum_{i \in \text{ range of } X} iP(X = i) + \sum_{j \in \text{ range of } Y} jP(Y = j).$$

Thus for independent[6] $X$ and $Y$

(6)  $E(X+Y) = E(X)+E(Y)$.

We can also show that for independent $X$ and $Y$

(7)  $E(XY) = E(X)E(Y)$,

(8)  $\sigma_{X+Y}^2 = E[\{X+Y-E(X+Y)\}^2] = \sigma_X^2 + \sigma_Y^2$,

and

(9)  $E[\{X-E(X)\} \{Y-E(Y)\}] = 0$.

The integral calculus can be invoked to show that (6), (7), (8), and (9) also hold for any pair $X$, $Y$ of independent, continuous random variables when the integrals in question exist.

Definition (2) can be extended in an obvious way to the following: $X_1$, $X_2$, ..., $X_n$ are *independent* random variables if

(10)  $F_{X_1, X_2,..., X_n}(\lambda_1, \lambda_2, ..., \lambda_n) = F_{X_1}(\lambda_1)F_{X_2}(\lambda_2) ... F_{X_n}(\lambda_n)$.

We can show in general that the following analogues of expressions (6)–(9) hold for independent random variables $X_1, ... , X_n$ and constants $a_1, ..., a_n$ and $b$:

(11)  $E \left( b + \sum_{i=1}^{n} a_i X_i \right) = b + \sum_{i=1}^{n} a_i E(X_i)$,

(12)  $E(X_1 X_2 ... X_n) = E(X_1)E(X_2) ... E(X_n)$,

(12a)  $E(X_i X_j) = E(X_i)E(X_j)$   $(i \neq j)$,

(13)  $\sigma^2_{\sum_{i=1}^{n} a_i X_i} = \sum_{i=1}^{n} a_i^2 \sigma_{X_i}^2$,

and

(14)  $E[\{X_k-E(X_k)\} \{X_l-E(X_l)\}] = 0$   $(k \neq l)$.

6   This assumption is actually not needed for sums of means. It is, however, necessary for sums of variances.

### 3.4.2 Standard error of proportion

A sequence of Bernoulli trials can be treated very elegantly in terms of joint distributions. In section 3.1, we learned that a sequence of $n$ Bernoulli trials resulted from the repetition of an experiment $\mathscr{E}$ with two outcomes, success and failure, where each replication of the experiment is independent of all the others and the probability of success at each trial is a fixed value $p$.

To treat this situation in terms of joint distribution, let

$$X_i = \begin{cases} 1 & \text{if the outcome of the } i\text{th trial is success,} \\ 0 & \text{if it is failure,} \end{cases}$$

for $i = 1, 2, \ldots, n$.

Note that at each trial there are only two possible outcomes $X_i = 0$ and $X_i = 1$, and since $P(X_i = 1) = p$ for all $i$, we must have

$$P(X_i = 0) = 1-p.$$

The probability of a sequence of outcomes $(e_1, e_2, \ldots, e_n)$ where $e_i$ is either success or failure is the probability of the compound event

$$(X_1 = e_1, X_2 = e_2, \ldots, X_n = e_n) = \bigcap_{i=1}^{n} (X_i = e_i).$$

Since each of the choices is independent of the others,

$$P(X_1 = e_1, X_2 = e_2, \ldots, X_n = e_n) = P(X_1 = e_1)P(X_2 = e_2)\ldots P(X_n = e_n),$$

and by (11) and (13), with $b = 0$ and $a_i = 1/n$ each time,

$$(15) \quad E\left(\frac{1}{n}\sum_{i=1}^{n} X_i\right) = \frac{1}{n}\sum_{i=1}^{n} E(X_i)$$

and

$$(16) \quad \sigma^2_{\frac{1}{n}\sum_{i=1}^{n} X_i} = \frac{1}{n^2}\sum_{i=1}^{n} \sigma_{X_i}^2.$$

To see what $E(X_i)$ and $\sigma_{X_i}^2$ equal, note that for our present choice of $X_i$ we have

$$E(X_i) = \sum_{j=0}^{1} jP(X_i = j) = (0 \cdot P(X_i = 0) + 1 \cdot P(X_i = 1)) = p$$

and

$$E[\{X_i - E(X_i)\}^2] = \sum_{j=0}^{1} \{j - E(X_i)\}^2 P(X_i = j)$$

$$= (0-p)^2 P(X_i = 0) + (1-p)^2 P(X_i = 1)$$

$$= p^2(1-p) + (1-p)^2 p = p(1-p).$$

Since

$$\bar{X}_n = \frac{1}{n} \sum_{i=1}^{n} X_i$$

is the average number of successes in a sample of size $n$, the expected average is given by (15) as

$$E(\bar{X}_n) = \frac{1}{n} \sum_{i=1}^{n} E(X_i) = \frac{1}{n} \sum_{i=1}^{n} p = p$$

and the variance of $\bar{X}_n$ is given by (16) as

$$\sigma_{\bar{X}_n}^2 = \frac{1}{n^2} \sum_{i=1}^{n} p(1-p) = \frac{p(1-p)}{n} .$$

The square root of $\sigma_{\bar{X}_n}^2$, denoted $\sigma_{\bar{X}_n}$, is called the *standard deviation* (or *error*) *of a proportion* for a sample of $n$ objects.

If $n$ is large it can be shown that, under the assumptions of this section, $\bar{X}_n$ is normally distributed with mean $\mu = E(\bar{X}_n)$ and variance $\sigma^2 = \sigma_{\bar{X}_n}^2$. Thus, for large $n$, we can use expression (20) of section 3.3 and a table of the normal distribution to show that

$$P(|\bar{X}_n - \mu| \leq \sigma) = P(|\bar{X}_n - \mu|/\sigma \leq 1)$$
$$= 2\{1 - N(1; 0, 1)\} = 0.6826.$$

Thus

(17) $P(|\bar{X}_n - p| \leq \sqrt{p(1-p)/n}) = 0.6826.$

Similarly, we can show that

$$P(|\bar{X}_n - p| \leq 2\sqrt{p(1-p)/n}) = 0.9544,$$

$$P(|X_n - p| \leq 1.645\sqrt{p(1-p)/n}) = 0.9000,$$

$$P(|X_n - p| \leq 1.960\sqrt{p(1-p)/n}) = 0.9500,$$

(18) $P(|X_n - p| \leq 2.579\sqrt{p(1-p)/n}) = 0.9900.$

EXAMPLE 3   Suppose we had reason to believe that a certain coin is not fair, i.e., that the probability of obtaining a head on a toss is not equal to the probability of obtaining a tail. We could test this using the above considerations. If the coin were fair, then a series of $n$ tosses could be taken to be a sequence of $n$ Bernoulli trials as indicated above. If we assume that $p = \frac{1}{2}$, then $E(X_i) = \frac{1}{2}$, and so for example

$$P(|\bar{X}_n - \tfrac{1}{2}| \leq 1.960/2\sqrt{n}) = P(|\bar{X}_n - \tfrac{1}{2}| \leq 0.980/\sqrt{n}) = 0.9500.$$

Thus

$$P(|\bar{X}_n - \tfrac{1}{2}| > 0.980\sqrt{n}) = 1 - P(|\bar{X}_n - \tfrac{1}{2}| \leq 0.980/\sqrt{n}) = 0.05.$$

If the value we obtained for $\bar{X}_n$, in a particular sample of $n$ tosses, differed from $\tfrac{1}{2}$ by more than $0.980/\sqrt{n}$, we would have reason to believe that the coin was biased. Indeed, the probability of obtaining such a deviation from the expected value of $\tfrac{1}{2}$ is less than 0.05 and could have happened in fewer than 5 cases in 100.

Example 3 is a special case of hypothesis testing which is the subject of chapter 5 to follow.

Often in the context of language study we find situations which can easily be discussed in terms of the model just considered. Suppose $U$ is a large set of, say, $m$ objects (perhaps a portion of running text or a random sample of words drawn from a text) in which $p$ is the proportion of objects with a certain property $\pi$. Suppose also that a sample of $n$ of these objects $e_1, e_2, ..., e_n$ is drawn from $U$ wherein each element of $U$ is as likely to be drawn as any other. Let

$$X_i = \begin{cases} 1 & \text{if } e_i \text{ has } \pi, \\ 0 & \text{if } e_i \text{ does not have } \pi, \end{cases}$$

for $i = 1, 2, ..., n$. Assume the choice of the $e_i$'s is *random*, i.e., each choice is independent of every other, and that $|U| = m$ is sufficiently greater than $n$ that the sampling without replacement does not change the probability $P(X_i = 1) = p$ significantly as $i$ varies from 1 to $n$.

Such a situation can be well approximated as a sequence of Bernoulli trials, in which case we can assume that $P(X_i = 1) = p$ and $P(X_i = 0) = 1 - p$ and proceed as indicated above.

In exercises 5 and 6 we discuss some of the applications of the section to language study, noting some of the pitfalls attendant upon assuming that the occurrences of successive events in running text are independent.

EXAMPLE 4    The random variable

$$Z = \sum_{i=1}^{n} X_i,$$

where the $X_i$ are defined as in the first displayed formula at the beginning of this subsection, is exactly the number of successes in a sequence of $n$ Bernoulli trials and so it has the binomial distribution

$$P(Z = k) = \binom{n}{k} p^k (1-p)^{n-k}$$

as we discovered in section 3.1. However, we can use the theory developed here to find $E(Z)$ and $\sigma_Z{}^2$ very easily. By (11) and (13)

$$E(Z) = \sum_{i=1}^{n} E(X_i) = np$$

and

$$\sigma_Z{}^2 = \sum_{i=1}^{n} \sigma_{X_i}{}^2 = np(1-p).$$

Compare the ease of this derivation to the work necessary to obtain $E(X)$ in section 3.2.

## EXERCISES

1 In example 1, find (a) $\sigma_{X+Y}{}^2$, (b) $E((X+Y)^2)$, (c) $E((X+Y)^3)$.

2 Verify that (6)–(9) hold for independent, non-negative, integer-valued random variables $X$, $Y$.

3 If $X_1, ..., X_n$ are independent, non-negative, integer-valued random variables, then

$$P(X_1 = i_1, X_2 = i_2, ..., X_n = i_n) = P(X_1 = i_1)P(X_2 = i_2)...P(X_n = i_n).$$

(a) Use this result to verify that (11)–(14) hold for such random variables.
(b) If $E(X_i) = \mu$ for $i = 1, 2, ..., n$, then show that $E(\bar{X}) = \mu$, where $\bar{X}$ is the random variable defined by

$$\bar{X} = \frac{1}{n} \sum_{i=1}^{n} X_i.$$

4 Let $X_1, X_2, ..., X_n$ each have the Poisson distribution with parameters $\lambda_1, \lambda_2, ..., \lambda_n$ respectively. If they are independent, what is the mean and variance of

$$\bar{X} = \frac{i}{n} \sum_{k=1}^{n} X_k?$$

(Try to prove this without assuming independence.)

5 Kučera and Francis (1967) found that the occurs in their test corpus with relative frequency 0.06899. Assume that the proportion of the's in any corpus is 0.06899. A sample of $n$ words is drawn at random from an arbitrary corpus, and the relative frequency $\bar{X}_n$ of the's is observed. By expression (18), the probability of obtaining an $\bar{X}_n$ that differs from 0.06899 by more than $2.579\sqrt{0.06899(1-0.06899)/n}$ is 0.01. Kučera and Francis (1967, p. 277) found that in a sample of some 162,173 words from 'learned and scientific writing' $\bar{X}_{100} = 7.73$. Is this evidence against the above assumption?

6 Note that the value given for the standard error of a proportion is computed on the assumption that the set of $X_i$'s is independent. Its validity can be called into question if this independence is not present. In various studies of vocabulary sampling such as

TABLE 3.4

| $x$ | Probability of $x$ occurring immediately following /ž/ in a corpus |
|---|---|
| /i/ | 0.1946 |
| /e/ | 0.4932 |
| /ə/ | 0.0008 |
| /a/ | 0.0154 |
| /o/ | 0.0236 |
| /g/ | 0.0015 |
| /y/ | 0.0150 |
| /h/ | 0.0014 |
| /r/ | 0.0039 |
| /w/ | 0.2507 |

those of D.W. Read (1949), Rebecca E. Hayden (1960), and A.H. Roberts (1965) the above standard error of a proportion is employed. In each of these studies the question of independence is ignored. Indeed independence is demonstrably missing. If we use Roberts's relative frequencies as estimates of probability, we obtain table 3.4 (Roberts 1965, p. 352). Presumably $x = $ /ž/ does not occur in the table because /žž/ did not occur in his sample.

In a sample of text containing $n$ phonemes, let

$$X_i = \begin{cases} 0 & \text{if the } i\text{th phoneme is /ž/,} \\ 1 & \text{if the } i\text{th phoneme is /i/,} \\ 2 & \text{if the } i\text{th phoneme is /e/,} \\ \cdots \\ 10 & \text{if the } i\text{th phoneme is /w/.} \end{cases}$$

(a) Use table 3.4 to show that $P(X_i = 0 \mid X_{i-1} = 0) = 0$.

(b) Find $P(X_i = k \mid X_{i-1} = 0)$ for $k = 1, ..., 10$. Compare these values with Roberts's estimates of $P(X_i = k)$ for $k = 1, 2, 3, 4, 5$ given in section 2.3. Does independence hold in these cases?

(c) From page 94 of Roberts we obtain $P(X_i = 0) = 0.0003$. From your experience of English or from part (a) argue that $P(X_i = 0, X_{i-1} = 0) = 0$ should hold. Find $P(X_i \neq 0 \mid X_{i-1} = 0)$ in table 3.4. How do you feel about the assumption of independence of the two events $X_i = 0$ and $X_{i-1} = 0$?

(d) From page 342 of Roberts we find that the relative frequency of /a/ after /a/ is 0.0009, so, using Roberts's relative frequencies as probabilities, we have $P(X_i = a \mid X_{i-1} = a) = 0.0009$. On the other hand from page 94 of Roberts we have $P(X_i = a) = 0.0463$. Find $P(X_i = a, X_{i-1} = a)$ and $P(X_i = a)P(X_{i-1} = a)$. How do you feel about the independence of the events $X_i = a$ and $X_{i-1} = a$ in this case?

(e) From pages 342 and 93 of Roberts we also find that we can say $P(X_i = ə \mid X_{i-1} = ə) = 0$ while $P(X_i = ə) = 0.1182$. How do you feel about the assumption of independence of the events $X_i = ə$ and $X_{i-1} = ə$?

(f) How do you feel about the reliability of the standard error of a proportion used on data of this kind? Under what circumstance would you expect the most reliable results in the application of the standard error of a proportion to phoneme counts?

## 3.5
### SOME LIMIT THEOREMS

In this section we state two results (without proof) which indicate why the arithmetic mean and the normal distribution play such a fundamental role in statistics and the theory of probability. Although the proofs of these results lie beyond the scope of this book, they are used implicitly, and often unconsciously, by most researchers who employ statistics.

Before stating the results, some definitions are needed. A sequence $X_1$, $X_2$, $X_3$, ... of random variables is said to be *identically distributed* if

$$F_{X_i}(x) = F_{X_j}(x)$$

for all real numbers $x$ and each pair $i, j$ selected from $N = \{1, 2, ...\}$. A sequence $X_1, X_2, X_3, ...$ of random variables is said to be *independent* if for every choice of a finite set of real numbers

$$\lambda_1, \lambda_2, \lambda_3, ..., \lambda_n$$

and a corresponding choice of distinct natural numbers

$$i_1, i_2, i_3, ..., i_n,$$

we have

$$P(X_{i_1} \leq \lambda_1, X_{i_2} \leq \lambda_2, ..., X_{i_n} \leq \lambda_n)$$
$$= P(X_{i_1} \leq \lambda_1) \cdot P(X_{i_2} \leq \lambda_2) \cdot ... \cdot P(X_{i_n} \leq \lambda_n).$$

One of the most useful examples of an identically distributed independent sequence of random variables has already been considered.

EXAMPLE 1  In section 3.4.2 the sequences of Bernoulli trials were considered in terms of the random variables $X_i$ equal to 1 if the $i$th trial is a success and equal to 0 if it is a failure. The sequence $X_1$, $X_2$, ... of these random variables clearly has the property that for any set of real numbers, $\lambda_1, \lambda_2, ..., \lambda_n$ and any corresponding set of distinct natural numbers $i_1, i_2, ..., i_n$,

$$P(X_{i_1} = \lambda_1, X_{i_2} = \lambda_2, ..., X_{i_n} = \lambda_n) = P(X_{i_1} = \lambda_1)P(X_{i_2} = \lambda_2)...P(X_{i_n} = \lambda_n).$$

Thus this sequence $X_1$, $X_2$, ... is such an independently distributed sequence of identically distributed random variables.

The first of the two results can now be stated.

THEOREM 1 (Law of Large Numbers)  Let $X_1, X_2, ...$ be a sequence of identically distributed independent random variables. Assume that their common mean and variance both exist and equal $\mu$ and $\sigma^2$ respectively. If

(1) $\bar{X}_n = \dfrac{1}{n} \sum\limits_{i=1}^{n} X_i,$

the arithmetic mean of the first $n$ $X_i$'s, then

(2) $P(|\bar{X}_n - \mu| \geq \varepsilon) \leq \sigma^2/n\varepsilon^2$

for every choice of a real number $\varepsilon > 0$.

EXAMPLE 2  Consider the Bernoulli trials of example 1. The mean of $X_j$ was shown in section 3.4.2 to be

$\mu = E(X_j) = p,$

and the variance was shown to be

$\sigma_{X_j}^2 = p(1-p).$

By theorem 1, we have

(3) $P(|\bar{X}_n - p| \geq \varepsilon) \leq \dfrac{p(1-p)}{n\varepsilon^2}.$

Since $p(1-p) \leq 1/4$ (see exercise 2 below),

$1(1-p)/n\varepsilon^2 \leq 1/4n\varepsilon^2.$

Thus

(4) $P(|\bar{X}_n - p| \geq \varepsilon) \leq 1/4n\varepsilon^2.$

Expression (4) can be used to estimate $p$ as follows. Suppose we wish to find $p$ to within 0.01 of its value. Expression (4) yields

$P(|\bar{X}_n - p| \geq 0.01) \leq 10^4/4n.$

Thus, whenever we make, say, $n = 10^5$ trials,

$P(|\bar{X}_n - p| \geq 0.01) \leq 1/40 = 0.025.$

This means that in not more than 75 cases out of 1000 will the value we obtain for $\bar{X}_{10^5}$ differ from the actual value of $p$ by more than 1/100, i.e., in more than 925 cases out of 1000,

$|\bar{X}_{10^5} - p| < 0.01.$

Put another way, in 925 cases out of 1000 the random interval $(\bar{X}_{10^5} - 0.01, \bar{X}_{10^5} + 0.01)$ contains the unknown constant $p$, i.e.,

$\bar{X}_{10^5} - 0.01 < p < \bar{X}_{10^5} + 0.01.$

EXAMPLE 3   Let us again consider Roberts's relative frequency of /ž/ (example 3 in section 3.1). Suppose we have a text of $n$ phonemes

$$\pi_1 \pi_2 \, ... \, \pi_n$$

in which we assume for the moment that the successive occurrences of /ž/ can be considered as the result of a sequence of Bernoulli trials, so that we can write

(5)   $P(\pi_1 = \check{z}, \pi_2 = \check{z}, ..., \pi_n = \check{z}) = P(\pi_1 = \check{z}) \cdot P(\pi_2 = \check{z}) \cdot ... \cdot P(\pi_n = \check{z})$,

and suppose that $P(\pi_i = \check{z}) = p$ for all $i = 1, 2, ..., n$. (The validity of these assumptions has been considered in exercise 6 of section 3.4.) If we let

$$X_i = \begin{cases} 0 & \text{if } \pi_i \neq /\check{z}/, \\ 1 & \text{if } \pi_i = /\check{z}/, \end{cases}$$

then the conditions of theorem 1 and example 2 are satisfied. Thus for a sample the size of Roberts's $n = 56{,}056{,}231$ phonemes, with $\bar{X}_n = 0.0003$, expression (4) becomes

(6)   $P(|0.0003 - p| \geq 0.001) \leq \dfrac{10^6}{4(56.056) \times 10^6} = 0.0045.$

Hence the probability that actually $0 \leq p < 0.0013$, to all intents and purposes, is one. This is, nevertheless, not a particularly good estimate. If we try for a better one and use $\varepsilon = 0.0001$, we obtain the following result from (4):

(6a) $P(|0.0003 - p| \geq 0.0001) \leq \dfrac{10^8}{4(56) \times 10^6} \approx 0.5.$

The most we can deduce from (6a) is that there is more than a 50–50 chance that

0.0002 $< p <$ 0.0004.

Suppose, however, we make an a priori assumption that the probability of /ž/'s occurrence is less than 1/100, (not unreasonable in the light of (6)); then (3) yields

$$P(|0.0003 - p| \geq 0.0001) \leq \frac{10^8(0.0099)}{(56) \times 10^6} \approx \frac{1}{56} = 0.018$$

i.e., in roughly 98 cases out of 100

0.0002 $< p <$ 0.0004.

For the higher-frequency phonemes, the hypothesis of independence seems to be less tenable than for /ž/ and so the use of (4) as a test for the reliability of an estimate of $p$ in these cases is less apt.

EXAMPLE 4   In sampling the number of articles in a hundred 50-word passages from *The Solid Mandala* by Patrick White, we obtained a sample mean $\bar{X}_{100} = 3.060$ with sample variance $s_{X_{100}}{}^2 = 3.007$. Assuming $s_{X_{100}}{}^2 = \sigma_X{}^2$, theorem 1 yields the result

$$P(|3.060 - \mu| \geqq 0.5) \leqq 3.007/100(0.25) = 0.1203.$$

Therefore with probability 0.8797

$$2.56 < \mu < 3.56.$$

The second of the results mentioned at the beginning of the section indicates the importance of the normal distribution.

THEOREM 2 (Central Limit Theorem)  Let $X_1$, $X_2$, ... be a sequence of independent identically distributed random variables with finite common mean $\mu$ and variance $\sigma^2$. If

$$Y_n = (X_1 + X_2 + \ldots + X_n - n\mu)/\sigma\sqrt{n},$$

then

$$(7) \quad F_{Y_n}(x) \to \frac{1}{\sqrt{2\pi}} \int_{-\infty}^{x} e^{-t^2/2}\, dt = N(x; 0, 1)$$

uniformly in $x$ as $n \to \infty$.

The word 'uniformly' in the enunciation of theorem 2 signifies that

$$\lim_{n \to \infty} F_{Y_n}(x) = N(x; 0, 1),$$

independent of the choice of $x$.

In the sense indicated in theorem 2, the normal distribution is the limiting distribution of $n\bar{X}_n$ when the independent $X_i$'s each have the same arbitrary distribution.

The power of this result is illustrated by the following corollary.

COROLLARY 1  Let $S_n$ be the number of successes in a sequence of $n$ Bernoulli trials, where $p$ denotes the probability of success in a single trial. If

$$Y_n = (S_n - np)/\sqrt{n(1-p)p},$$

then

$$F_{Y_n}(x) \to \frac{1}{\sqrt{2\pi}} \int_{-\infty}^{x} e^{-t^2/2}\, dt$$

uniformly in $x$ as $n \to \infty$.

We have already used this corollary in example 3 of section 3.3.

PROOF OF COROLLARY 1    In examples 1 and 2 above and in section 3.4.2, we found that if $X_i$ is the random variable defined at the beginning of section 3.4.2, then

$$S_n = X_1 + X_2 + \ldots + X_n,$$

$\mu = p$, and $\sigma^2 = p(1-p)$. By theorem 2, if we designate

$$Y_n = (S_n - np)/\sqrt{np(1-p)},$$

then expression (7) holds for this $Y_n$.

REMARK    As was noted earlier, it can be easily shown that the random variable $S_n$ has the binomial distribution, so that

$$P(S_n = k) = b(k; n, p) = \binom{n}{k} p^k (1-p)^{n-k}.$$

Therefore the proof of corollary 1 above establishes the truth of expression (15) in section 3.3, since an exercise in the integral calculus shows that

$$F_{S_n}(x) = N(x; np, \sqrt{np(1-p)})$$

if and only if

$$F_{S_n - np/\sqrt{np(1-p)}}(x) = N(x; 0, 1).$$

EXAMPLE 5    Using the data and assumptions of example 3, we can compute $Y_n$ as follows:

$$Y_n = (S_n - np)/\sqrt{n(1-p)p} = \sqrt{n}\left(\frac{S_n}{n} - p\right)\Big/\sqrt{(1-p)p}$$

$$\approx \sqrt{56 \times 10^3}(0.0003 - p)/\sqrt{3 \times 10^{-2}}$$

$$= \sqrt{56/3} \times 10^5 (0.0003 - p).$$

From a table of the normal distribution, we obtain

$$F_{Y_n}(3) = 0.9987 = P(Y_n \leq 3)$$

and

$$F_{Y_n}(-3) = 0.0013 = P(Y_n \leq -3)$$

so

$$P(|Y_n| \leq 3) = F_{Y_n}(3) - F_{Y_N}(-3) = 0.9974.$$

Thus in at least 99 cases in 100

$$|Y_n| = \sqrt{56/3}\,(10^5)\,|0.0003 - p| \leq 3$$

or

$$|0.0003 - p| \leqq \frac{3\sqrt{3}}{\sqrt{56}} (10^{-5}) < 10^{-5}.$$

Therefore

$$0.00029 < p < 0.00031$$

with probability 0.9974 under the assumptions of example 3. Since $np$ is very much larger than 5, the assumption of normality for $F_{Y_n}(x)$ results in a good approximation. (Cf. section 3.4.)

EXAMPLE 6   Using the data and assumptions of example 4, together with the assumption that $n = 100$ is large enough for the distribution of $Y_{100}$ to be well approximated by the normal distribution $N(x; 0, 1)$, we obtain

$$P(|Y_{100}| \leqq 3) = 0.9974$$

where

$$Y_{100} = (306 - 100\mu)/30.07 = (3.06 - \mu)/0.3007.$$

Thus

$$|3.06 - \mu| \leqq 3(0.3007) = 0.9021 \quad \text{or} \quad 2.158 \leqq \mu \leqq 3.962$$

with probability 0.9974.

$$P(|Y_{100}| \leqq 1) = 0.6827,$$

so

$$|3.06 - \mu| \leqq 0.3007 \quad \text{or} \quad 2.760 \leqq \mu \leqq 3.360$$

with probability 0.6827.

It can be seen from these examples that the central limit theorem is a potentially more powerful tool than the law of large numbers.

EXERCISES

1 In his phoneme count, A.H. Roberts (1965, p. 112) found that in 56,056,231 phonemes there were 20,240,466 vowels. Let $X_i$ be the random variable which equals 1 if the $i$th phoneme selected is a vowel and 0 otherwise.

(a) Find $\bar{X}_n = \sum_{i=1}^{56,056,231} X_i/56,056,231$ for this sample.

(b) Use expression (4) with $\varepsilon = 0.001$ to estimate the accuracy with which $\bar{X}_{56,056,231}$ approximates the actual probability of a vowel under the independence hypothesis implicit in (4).

(c) Use the central limit theorem to estimate the accuracy of $\bar{X}_{56,056,231}$. Is the use of this theorem really justifiable in this example?

2 Prove that if $p_1$ and $p_2$ are non-negative real numbers such that $p_1 + p_2 = 1$, then $p_1 p_2 \leqq 1/4$. (Hint: $p_2 = 1 - p_1$, and $p_1$ can be written $p_1 = \frac{1}{2} + \varepsilon$ for some $\varepsilon$ where $0 \leqq \varepsilon \leqq 1$. Form $p_1 p_2$ in terms of $\varepsilon$.)

# 4
# Estimation

The notions of sample and population were introduced in section 1.1. You will remember that a population is a set of objects to be observed. It may be a finite set like the set of inhabitants of a certain area, or a countably infinite set like the set of possible natural numbers corresponding to the first success in an infinite sequence of Bernoulli trials, or an uncountably infinite set like the set of points between 0 and 1, the possible values of a point selected at random from the interval [0, 1]. Our general interest has centred on the values assumed by variates (or random variables) defined on such populations. In chapters 2 and 3 we saw how a probability space could be defined for the experiment of selecting an individual from the population and measuring the value of a variate $X$ for that individual.

A sample from the population was taken (in section 1.1) to be an ordered collection $(I_1, I_2, ..., I_n)$ of individuals drawn from the population in a particular experiment. A sample of a variate $X$ defined over the population was assumed to be the $n$-tuple $(x_1, x_2, ..., x_n)$ such that for each $j = 1, 2, ..., n$, $x_j$ is the value that $X$ assumes for individual $I_j$.

The succession of values $(x_1, x_2, ..., x_n)$ may depend on the way the sample is chosen, so that the probability that $X$ takes on a certain value for the $j$th entry in a sample from the population may differ from the probability that $X$ takes on that value for the $(j+1)$st (e.g., in the case of sampling without replacement). Therefore it is sometimes useful not to think of the sample $(x_1, x_2, ..., x_n)$ as a sequence of values that a single variate $X$ may assume but rather to consider each entry $x_j$ separately as the value that a separate random variable $X_j$ can assume. Thus a sample $(x_1, x_2, ..., x_n)$ of size $n$ is often assumed to be the $n$-tuple of values taken on by the $n$-tuple of variates $(X_1, X_2, ..., X_n)$ as a result of the performance of a certain experiment. When taking this point of view, we say that $X_i$ is the random variable *corresponding* to entry $x_i$ in the sample. The following example illustrates why such a distinction is useful.

EXAMPLE 1  Consider an urn containing $n_1$ red balls and $n_2$ white balls. Let

$n \leq n_1 + n_2$ balls be selected without replacement from the urn. Let 0 and 1 stand respectively for the selection of a white and a red ball. Then the sample can be construed as the $n$-tuple $(x_1, x_2, ..., x_n)$ where $x_i$ equals 0 if the $i$th ball selected is white and 1 otherwise, and each $x_i$ in the sample $(x_1, ..., x_n)$ is a value of the random variable $X_i$ where

$$X_i = \begin{cases} 1 & \text{if the } i\text{th ball is red,} \\ 0 & \text{if the } i\text{th ball is white.} \end{cases}$$

Thus $(x_1, x_2, ..., x_n)$ is an $n$-tuple of values of the $n$-tuple $(X_1, X_2, ..., X_n)$ of random variables. Note that, because we are not replacing the balls each time we draw one out, the probability of $X_2$ being 1 depends on the value that $X_1$ assumed. Thus the random variables $X_1$ and $X_2$ are not independent. Indeed

$$P(X_2 = 1 \mid X_1 = 1) = \frac{n_1 - 1}{n_1 + n_2 - 1}$$

and

$$P(X_2 = 1 \mid X_1 = 0) = \frac{n_1}{n_1 + n_2 - 1}.$$

If after each selection the ball were replaced, we would have

$$P(X_2 = 1 \mid X_1 = 1) = P(X_2 = 1 \mid X_1 = 0) = \frac{n_1}{n_1 + n_2}$$

and

(1)  $P(X_1 = 1, X_2 = 1) = P(X_1 = 1)P(X_2 = 1),$

so the two events $X_1 = 1$ and $X_2 = 1$ would be independent.

In general a sample $(x_1, x_2, ..., x_n)$ is said to be a *random sample from a distribution* if (i) for each $i = 1, 2, ..., n$ the random variable $X_i$ corresponding to $x_i$ has the same distribution as all the other corresponding random variables, and (ii) the set $\{X_1, X_2, ..., X_n\}$ of corresponding random variables is independent.

Thus the random variables $X_1, X_2, ..., X_n$ corresponding to $x_1, x_2, ..., x_n$ respectively are independent and identically distributed as these terms are defined in section 3.5. From a statistical point of view there is no need *in this case* to distinguish the $X_i$'s, so we could, if we wished, return to thinking of the sample $(x_1, x_2, ..., x_n)$ as a sample of values of a single variate $X$ (with the common distribution of the $X_i$'s) that was introduced in section 1.1. It will be useful to be able to switch from one interpretation to the other, so we shall want to use both of them.

In example 1, if the sampling is done without replacement, then neither (i) nor (ii) is fulfilled, and so the sample $(x_1, x_2, ..., x_n)$ is not a random sample from a distribution. If, however, the sampling is done with replacement, then

$$(2) \quad P(X_i = 1) = \frac{n_1}{n_1 + n_2}, \qquad P(X_i = 0) = \frac{n_2}{n_1 + n_2}$$

for each $i = 1, 2, \ldots, n$, and it is easy to extend (1) to show that $\{X_1, X_2, \ldots, X_n\}$ is an independent set of random variables. Thus (i) and (ii) are satisfied in this case and the sample is a random sample from a distribution.

In what follows, we shall usually be considering random samples from a distribution, and when there is no possibility of confusion we shall simply refer to these as *random samples*.

Note that a sample $(x_1, x_2, \ldots, x_n)$ obtained from a running spoken-language text containing $n$ phonemes where

$$x_i = \begin{cases} 1 & \text{if the } i\text{th phoneme is } /\check{z}/, \\ 0 & \text{otherwise,} \end{cases}$$

is strictly speaking *not* a random sample because (ii) is not satisfied. Sometimes, as we have observed, if the dependence is not too marked, the assumption of independence yields a good approximation.

EXAMPLE 2   Let $S_0$ be the set of individuals in a certain geographical area of which an unknown proportion $p$ are native speakers of a particular language $L$. Suppose we draw a sample of $n$ speakers from the population in such a way that each subset of $n$ speakers is as likely to be drawn as any other and then note the number of native speakers of $L$. If the population is large enough compared to $n$, the sample $(x_1, x_2, \ldots, x_n)$, where $x_i$ is 1 or 0 according as individual $i$ has $L$ as his native language or not, can be assumed to be a random sample from the distribution

$$F_{X_i}(x) = \begin{cases} 0, & x < 0, \\ 1-p, & 0 \le x < 1, \\ 1, & 1 \le x, \end{cases}$$

where the probability that $X_i = 0$ is $1-p$ and the probability that $X_i = 1$ is $p$. Our main interest lies in finding the value of $p$, that is, we wish to *estimate* $p$.

In general, the problem of assessing the value of a parameter (like $p$ in this example) for a population on the basis of a sample drawn from that population is known as the problem of *estimation*.

In general, let us assume that our random variable $X$ has a distribution depending on a parameter $\theta$, so that the CDF of $X$ is

$$F_X(x) = F(x, \theta)$$

where $\theta$ is unknown to us. Suppose $(x_1, x_2, \ldots, x_n)$ is a random sample from the distribution of $X$; then we can assume for each $i$ that $x_i$ is a value of a random variable $X_i$ which has the same distribution as $X$ and that the $X_i$ are independent. We might take some random variable which is a function $T_n(X_1, X_2, ..., X_n)$ of the $X_i$ as our estimate of the parameter $\theta$. Such an estimating random variable as $T_n$ is sometimes called a *statistic*. The natural question to ask is: When is this statistic $T_n$ a good estimate of $\theta$? Naturally, we would like $T_n$ to be close to $\theta$ in some sense.

There are two such senses that we shall consider here. An estimate $T_n$ of a parameter $\theta$ is *unbiased* when $E(T_n) = \theta$, and the estimate $T_n$ is a *consistent* estimate of $\theta$ if, as the sample size $n$ increases, the probability

$$P(|T_n - \theta| \geq \varepsilon) \to 0$$

for every choice of $\varepsilon > 0$.

Returning to example 2, let $(x_1, x_2, ..., x_n)$ be a random sample from the common distribution of the $X_i$. Since $E(X_i) = 0(1-p) + 1p = p$, it follows that the function

$$T_n = \bar{X}_n = \frac{1}{n} \sum_{i=1}^{n} X_i,$$

whose value for the sample $(x_1, x_2, ..., x_n)$ is

$$\bar{x} = \frac{1}{n} \sum_{i=1}^{n} x_i,$$

constitutes an unbiased estimate of $p$. Indeed, by the development in section 3.4.2, $E(\bar{X}_n) = p$.

In general, the law of large numbers states that, under the hypothesis of theorem 1 in section 3.5,

(3) $\quad P(|\bar{X}_n - \mu| \geq \varepsilon) \leq \sigma^2/n\varepsilon^2$

where $\mu$ and $\sigma^2$ are the mean and variance of the common distribution of the $X_i$. Since when $\sigma$ exists the right side of the inequality in (3) approaches 0 as $n \to \infty$, $\bar{X}_n$ is a consistent estimate of $\mu$ whenever both $\mu$ and $\sigma$ exist.

In example 2, the mean $\mu = p$ and the variance (see section 3.4.2)

$$\sigma^2(X_n) = \sigma_{X_n}^2 = p(1-p)$$

both exist, so that the hypothesis of theorem 1 in section 3.5 is satisfied and (3) becomes

$$P(|\bar{X}_n - p| \geq \varepsilon) \leq p(1-p)/n\varepsilon^2.$$

Therefore $\bar{X}_n$ is a consistent estimate of $p$.

In the present chapter, our purpose is to review two of the methods used for estimating the parameters of distributions.

## EXERCISES

1 As we have seen in chapter 2, the occurrence of articles in 50-word passages in certain literary genres appears to follow the Poisson distribution so that if $X$ is the number of articles in a 50-word passage,

$$P(X = x) = \lambda^x e^{-\lambda}/x!$$

for $x = 0, 1, 2, \ldots$ Suppose that in a random sample $(P_1, P_2, \ldots, P_n)$ of 50-word passages taken from a certain novel there are $x_i$ articles in passage $P_i$ for $i = 1, 2, \ldots, n$. Show that

$$\hat{\lambda} = \frac{1}{n} \sum_{i=1}^{n} x_i$$

is a consistent unbiased estimate of $\lambda$.

2 Suppose that a random variable $X$ is known to follow the normal distribution, but the parameter $\mu$ is unknown; suppose a value of $X$ is obtained by performing a certain experiment $\mathscr{E}$; and finally suppose this experiment is performed $n$ times and each time $X$ is observed, resulting in a sample $(x_1, x_2, \ldots, x_n)$ where $x_i$ is the value of $X$ obtained for the $i$ replication of $\mathscr{E}$. Use the law of large numbers to obtain a consistent estimate of $\mu$ and show that it is unbiased.

## 4.2

### THE METHOD OF MOMENTS

The first method of estimating parameters which we shall discuss is by far the easiest one to apply. The *method of moments* rests on the law of large numbers. For example, consider the $k$th sample moment

$$m'_{k,n} = \frac{1}{n} \sum_{i=1}^{n} x_i^k$$

for a random sample $(x_1, x_2, \ldots, x_n)$ drawn from the distribution of a variate $X$. $m'_{k,n}$ can be thought of as the value of a random variable

(1) $\quad M'_{k,n} = \frac{1}{n} \sum_{i=1}^{n} X_i^k$

where the $X_i^k (i = 1, 2, \ldots, n)$ form a set of independent identically distributed random variables. If the $X_i^k$ possess a finite common mean and variance (which is true in most cases we need to consider), then by the law of large numbers

(2) $\quad P(|M'_{k,n} - E(X_i^k)| \geq \varepsilon) \leq \sigma^2(X_i^k)/n\varepsilon^2$

and so

(3) $\quad \lim_{n \to \infty} P(|M'_{k,n} - E(X_i^k)| \geq \varepsilon) = 0$

for all $\varepsilon > 0$. Thus the value $m'_{k,n}$ of $M'_{k,n}$ for any sample $(x_1, x_2, \ldots, x_n)$ will differ from $E(X_i^k)$, the $k$th moment of the common distribution of the $X_i$, by more than $\varepsilon$ with probability less than $\sigma^2(X_i^k)/n\varepsilon^2$, i.e., less than some constant times $1/n$. Thus, in particular, $m'_{k,n}$ is a consistent estimate of the $k$th moment about the origin of the common distribution of the $X_i$.

In general, if for a random sample $(x_1, x_2, \ldots, x_n)$, $m'_{r,n}$ stands for the $r$th sample moment about the origin (see section 1.6), $m_{r,n}$ stands for the $r$th sample moment about the sample mean $m'_{1,n}$ (see also section 1.6), and $\mu_r'$ and $\mu_r$ stand for the corresponding distributional moments (see section 3.2), then it can be proved that

(4)  $E(m'_{r,n}) = \mu_r'$,

(5)  $\sigma^2(m'_{r,n}) = \dfrac{1}{n}[\mu_{2r}' - (\mu_r')^2]$,

(6)  $\lim_{n \to \infty} E(m_{r,n}) = \mu_r$,

(7)  $\lim_{n \to \infty} \sigma^2(m_{r,n}) = \dfrac{1}{n}(\mu_{2r} - \mu_r^2 + r^2\mu_{r-1}\mu_2 - 2r\mu_{r-1}\mu_{r+1})$.

Thus, by (4) the sample moments about the origin are not only consistent but also unbiased estimates of the corresponding distribution moments. However, because (6) is all that can be proved for the central sample moments, these are only unbiased asymptotically[1] for large $n$. Consistency can be shown to hold in both cases.

*The method of moments consists in using the sample moments as estimates of the corresponding distribution moments.* Thus, if a variate $X$ is known to have a distribution of the form

$$F_X(x) = F(x; \rho_1, \rho_2, \ldots, \rho_k)$$

depending on $k$ parameters, then to estimate the parameters $\rho_1, \rho_2, \ldots, \rho_k$ using the method of moments we can proceed as follows: (i) Express each $\rho_i$ as a function of distribution moments. (ii) Take a sample and compute from it the sample moments corresponding to the distribution moments mentioned in (i). (iii) Form the estimate $\hat{\rho}_i$ by replacing the distribution moments in its expression by the numerical values of the corresponding sample moments obtained in (ii). This procedure will result in consistent asymptotically unbiased estimates of the $\rho_i$.

To verify that central moments may indeed be biased for small $n$, consider the sample variance $m_{2,n}$ constructed from a random sample $(x_1, x_2, \ldots, x_n)$ from a distribution which is known to have finite mean and variance. We can

---

1  A statistic $T_n(X_1, \ldots, X_n)$ is an asymptotically unbiased estimate of a parameter $\theta$ provided $\lim_{n \to \infty} E(T_n(X_1, \ldots, X_n)) = \theta$, i.e., the expected value of the statistic approaches the limit $\theta$ as the sample size $n$ becomes large.

think of $x_i$ as the value of a corresponding random variable $X_i$ which has the distribution in question. Then

$$E(s^2) = E(m_{2,n}) = E\left(\frac{1}{n}\sum_{i=1}^{n}\left[X_i - \frac{1}{n}\sum X_i\right]^2\right)$$

$$= \frac{1}{n}E\left(\sum_{i=1}^{n}X_i^2 - \frac{1}{n}\sum_{i=1}^{n}\sum_{j=1}^{n}X_iX_j\right)$$

$$= \frac{1}{n}\left(\sum_{i=1}^{n}E(X_i^2) - \frac{1}{n}\sum_{i=1}^{n}\sum_{j=1}^{n}E(X_iX_j)\right).$$

Since the sample is random, the set $X_1, X_2, ..., X_n$ of random variables is independent, so by (12a) in section 3.4

$$E(X_iX_j) = E(X_i)E(X_j)$$

for $i \neq j$. Therefore

$$E(m_{2,n}) = \frac{1}{n}\left(\sum_{i=1}^{n}E(X_i^2) - \frac{1}{n}\sum_{i=1}^{n}E(X_i^2) - \frac{1}{n}\sum_{i\neq j}E(X_i)E(X_j)\right)$$

$$= \frac{1}{n}\left(n\mu_2' - \mu_2' - \frac{1}{n}(n-1)\,n\,(\mu_1')^2\right)$$

$$= \frac{n-1}{n}[\mu_2' - (\mu_1')^2] = \frac{n-1}{n}\mu_2 = \frac{n-1}{n}\sigma^2.$$

Therefore we can say only that $E(m_{2,n}) = \mu_2$ asymptotically for large $n$, and so $s^2 = m_{2,n}$ is a biased estimator of $\mu_2 = \sigma^2$ in general.

In order to render the sample variance unbiased it is usually multiplied by $n/(n-1)$ to obtain

$$(8) \quad S^2 = \frac{1}{n-1}\sum_{i=1}^{n}(x_i - \bar{x})^2 = \frac{n}{n-1}s^2,$$

the so-called *unbiased sample variance*. In most studies this unbiased sample variance is used without comment in place of the (biased) sample variance defined in chapter 1.

To see that $S^2$ is indeed an unbiased estimate of $\sigma^2$, note that

$$E(S^2) = E\left(\frac{n}{n-1}s^2\right) = \frac{n}{n-1}E(m_{2,n}) = \frac{n}{n-1}\frac{n-1}{n}\sigma^2 = \sigma^2.$$

Since

$$\lim_{n\to\infty}\frac{n}{n-1} = 1,$$

the unbiased sample variance $S^2$ is also a consistent estimate of the population variance.

For the remainder of the book the term sample variance, still designated $s^2$, refers to the *unbiased sample variance* unless otherwise specified.

Now let us use the method of moments to estimate a few parameters. In exercise 1 of section 4.1 we saw how the law of large numbers could be used to obtain an estimate of $\lambda$ for a Poisson variate. Consider now a random variable $X$ with binomial distribution $b(x; n, p)$ where $n$ is known but $p$ is not. Using the method of moments, the distribution mean of $X$ is $E(X) = np$. For a random sample of $X$, say $(x_1, x_2, ..., x_n)$, the sample mean is a consistent unbiased estimate of $E(X)$. Thus we have

$$\hat{p} = \frac{1}{n}\bar{X} = \frac{1}{n^2}\sum_{i=1}^{n} x_i,$$

which is the consistent unbiased estimate of $p$ furnished by the method of moments.

For the negative binomial distribution $n(x; k, p)$, we found (see section 3.3) that

$$E(X) = k(1-p)/p \quad \text{and} \quad \sigma_X{}^2 = k(1-p)/p^2.$$

From the method of moments it follows that if $(x_1, x_2, ..., x_n)$ is a random sample of a variate $X$ with negative binomial distribution where both $k$ and $p$ are unknown, then the sample mean $\bar{X}$ furnishes a consistent unbiased estimate of $E(X)$, and

$$s^2 = \frac{1}{n-1}\sum_{i=1}^{n} (x_i - \bar{X})^2$$

furnishes a consistent unbiased estimate of $\sigma_X{}^2$. Thus the estimates $\hat{k}$ and $\hat{p}$ of $k$ and $p$ should be so chosen that

(9) $\bar{X} = \hat{k}(1-\hat{p})/\hat{p}$

and

(10) $s_X{}^2 = \hat{k}(1-\hat{p})/\hat{p}^2$.

Therefore

(11) $\hat{p} = \bar{X}/s^2$

and

(12) $\hat{k} = \hat{p}\bar{X}/(1-\hat{p})$.

In some cases the method of moments yields more than one estimate of a parameter. For example, if $X$ has the Poisson distribution so that

$$P(X = x) = \frac{\lambda^x}{x!} e^{-\lambda}$$

for $x = 0, 1, 2, \ldots$, then it is a simple matter to show that if $(x_1, x_2, \ldots, x_n)$ is a random sample from the distribution of $X$

$$\lambda_0 = \frac{1}{n} \sum_{j=1}^{n} x_j = \bar{X}$$

and

$$\lambda_1 = s^2 = \frac{1}{n-1} \sum_{j=1}^{n} (x_j - \bar{X})^2$$

are both consistent unbiased estimates of $\lambda$. Which one is more accurate? If we consider $\lambda_i$ as the value of the random variable $\Lambda_i$ $(i = 0, 1)$ computed from the sample $(x_1, x_2, \ldots, x_n)$, then the more accurate estimate will, of course, correspond to the random variable with the smaller variance. The variance of $\Lambda_0$ can be computed from equation (5):

$$\sigma^2(\Lambda_0) = \sigma^2 \left( \frac{1}{n} \sum_{i=1}^{n} X_i \right) = \frac{1}{n} \lambda.$$

The variance of $\Lambda_1$ can be computed with a little more difficulty from equation (7):

$$\sigma^2(\Lambda_1) = \sigma^2 \left( \frac{n}{n-1} m_{2,n} \right) = \frac{n^2}{(n-1)^2} \sigma^2(m_{2,n})$$

$$\approx \frac{n}{(n-1)^2} (\mu_4 - \mu_2{}^2 + 4\mu_1\mu_2 - 4\mu_1\mu_3).$$

Since $\mu_1$ is the first central moment, it is zero, and since $\mu_2 = \sigma^2 = \lambda$,

$$\sigma^2(\Lambda_1) \approx \frac{n}{(n-1)^2} (\mu_4 - \lambda^2).$$

To find $\mu_4$, we must evaluate the series

$$\mu_4 = \sum_{j=0}^{\infty} \frac{(x - \lambda)^4 \lambda^x}{x!} e^{-\lambda}.$$

A rather lengthy and unpleasant calculation yields the result

$$\mu_4 = 3\lambda^2 + \lambda,$$

and so

$$\sigma^2(\Lambda_1) = \frac{n}{(n-1)^2} (2\lambda^2 + \lambda).$$

Therefore, for large enough $n$, we can assume that

$$\sigma^2(\Lambda_1) > \sigma^2(\Lambda_0).$$

Hence the sample mean is a better estimate of $\lambda$ than the sample variance.

## EXERCISES

1 Use the method of moments to obtain two estimates for $p$ based on a random sample $(x_1, x_2, ..., x_n)$ of a binomially distributed random variable $X$ where the number of trials is known.

2 For each of the estimates $p_0$ and $p_1$ obtained in exercise 1 use the law of large numbers to ascertain the reliability of the estimate in each case. (Hint: $\mu_2' = n(n-1)p^2 + np$, $\mu_3' = n(n-1)(n-2)p^3 + 3\mu_2' - 2np$, and $\mu_4' = n(n-1)(n-2)(n-3)p^4 + 6\mu_3' - 11\mu_2' + 6np$.)

3 Suppose a variate $X$ can take on values 0, 1, 2, ..., $m$ with probability

$$P(X = k) = \binom{m}{k} (\tfrac{1}{2})^m$$

for $k = 0, 1, 2, ..., m$. Suppose also that we do not know $m$. However, we have a random sample $(x_1, x_2, ..., x_n)$ of values of $X$.
   (a) Find $E(X)$ and use the method of moments to estimate $m$.
   (b) Use the random sample $(1, 1, 1, 1, 11)$ of values of $X$ to show that the method of moments does not always yield a reasonable estimate of the parameters in question. (Hint: since $X$ can equal 11 in this case, it could not have been the case that $P(X = 11) = 0$. Therefore $m \geq 11$ must hold.)

4 Suppose $w_1w_2 ... w_n$ is an arbitrary $n$-word piece of running text from some literary work where each $w_i$ stands for a word and the punctuation has been removed. Consider a set $K$ of word-types and assume that the number of instances of words from $K$ in $w_1w_2 ... w_n$ follows the binomial distribution $b(x; m, p)$ with $m$ and $p$ unknown. Use the method of moments to estimate $m$ and $p$. (Hint: consider a sample of $k$ such running texts with $x_1, x_2, ..., x_k$ words from $K$ in each case, and follow the procedures (i), (ii), and (iii) above.)

5 In a sampling of fifty 50-word passages drawn at random from *Victory* by Joseph Conrad, the number of articles was counted. It was found that the sample mean and unbiased sample variance of $X$, the number of articles per passage, were $\bar{X} = 3.86$ and $s_x^2 = 2.53$ respectively. Using the model given in exercise 4 above, estimate $m$ and $p$ with this sample.

6 In a sample of fifty 50-word passages drawn at random from *The Wapshot Chronicle* by John Cheever the average number of articles was $\bar{X} = 4.52$ and the unbiased sample variance was $s_x^2 = 4.54$. What can you say about the appropriateness of the binomial model mooted in exercise 4 for this case?

4.3

MAXIMUM LIKELIHOOD ESTIMATES

Sometimes estimates of the parameters of a distribution using the method of moments are either not available or not satisfactory. In such cases, the next estimation technique to which one usually resorts is the method of maximum likelihood developed by R.A. Fisher.

Consider again example 2 in section 4.1. There we wished to obtain an estimate of the probability $p$ that a person selected at random from a certain population was a native speaker of $L$. We chose a random sample of individuals $(I_1, I_2, ..., I_n)$ from the population to obtain a sample $(x_1, x_2, ..., x_n)$ of the random variable

$$X(I) = \begin{cases} 1 & \text{if } I \text{ is a native speaker of } L, \\ 0 & \text{otherwise.} \end{cases}$$

Thus, for each $i$, $x_i = 1$ if $I_i$ is a native speaker of $L$ and $x_i = 0$ otherwise. As an alternative to the method of moments, we might proceed to find a 'good' estimate of $p$ as follows. Let us conceive of $X_i$ as the random variable associated with the $i$th observation in our sample. $X_i$ and $X_j$ both have the same distribution, and $X_i$ and $X_j$ are independent because the sample is random. Then

$$P(X_i = 1) = p \quad \text{and} \quad P(X_i = 0) = q = 1-p.$$

Since the sample is random, the probability of obtaining exactly the sample $(x_1, x_2, ..., x_n)$ is given by the expression

$$(1) \quad L(x_1, x_2, ..., x_n; p) = P(X_1 = x_1, X_2 = x_2, ..., X_n = x_n)$$
$$= P(X_1 = x_1)P(X_2 = x_2)...P(X_n = x_n)$$
$$= p^{\sum_{i=1}^{n} x_i}(1-p)^{n - \sum_{i=1}^{n} x_i}$$

provided, of course, that $p$ is known. The function $L(x_1, x_2, ..., x_n; p)$ of $p$ is called the *likelihood of the random sample* $(x_1, x_2, ..., x_n)$ and depends on the value of $p$. Since this particular sample was actually obtained, it might seem that a good estimate of $p$ would be that one which, if it exists, renders the likelihood (probability) of this sample largest.

Since $n$, $x_1, x_2, ..., x_n$ are fixed constants, $L$ is a function of $p$. Our problem is to find that value, or those values if there are more than one, of $p$ which make $L$ maximum. One notes first that $L$ is a continuous function of $p$ in the domain $[0, 1]$, and that

$$L(x_1, x_2, ..., x_n; p) \geqq 0$$

for all $p \in [0, 1]$. In addition,

$$L(x_1, x_2, ..., x_n; 0) = L(x_1, x_2, ..., x_n; 1) = 0,$$

so the values of $p$ which yield the maximum values for $L$ must lie in the open interval

$$(0, 1) = \{p \mid 0 < p < 1\}.$$

One of the results of the calculus[2] can be used to find the values of $p$ corresponding to $L$ maximum. In cases like these, if $L$ has a maximum for some $p_0$, then

$$\frac{dL}{dp} = 0$$

for that $p_0$.

The computation involved in finding this derivative can often be lessened by noting that for a continuous non-vanishing function $f(x)$

(2) $\dfrac{df(x)}{dx} = 0$

if and only if

(3) $\dfrac{d \ln f(x)}{dx} = \dfrac{1}{f(x)} \dfrac{df(x)}{dx} = 0,$

where $\ln x$ is the natural logarithm of $x$, i.e., $\ln x = y$ if and only if $e^y = x$. In our case

$$\ln L = \left( \sum_{i=1}^{n} x_i \right) \ln p + \left( n - \sum_{i=1}^{n} x_i \right) \ln(1 - p),$$

and so the maximum occurs for some value of $p$ for which

$$0 = \frac{d(\ln L)}{dp} = \frac{1}{p} \left( \sum_{i=1}^{n} x_i \right) + \frac{-1}{1-p} \left( n - \sum_{i=1}^{n} x_i \right)$$

$$= \left[ (1-p) \sum_{i=1}^{n} x_i - pn + p \left( \sum_{i=1}^{n} x_i \right) \right] \frac{1}{p(1-p)},$$

i.e., for which

(4) $\left( \displaystyle\sum_{i=1}^{n} x_i \right) - pn = 0.$

This yields a unique maximum likelihood estimate

(5) $\hat{p} = \dfrac{1}{n} \displaystyle\sum_{i=1}^{n} x_i.$

2  Those without knowledge of the calculus can skip the technical details and go directly to the result, in this case expression (5).

A check of the values of $L$ near $\frac{1}{n} \sum_{i=1}^{n} x_i$ will indicate that the value of $p$ in (5) indeed renders $L$ maximum.

Note that the estimate given by (5) is identical with that obtained by the method of moments. Thus it is possible that the results of this method will not differ from the estimates obtained by the method of moments.

In general, if we know the common distribution of the random variable $X_i$ corresponding to the $x_i$ in the random sample $(x_1, x_2, ..., x_n)$ up to some fixed parameter $\theta$ associated with that distribution, then the following procedure yields the *maximum likelihood estimate of* $\theta$.

(i) Form the likelihood function: either

(6)  $L(x_1, ..., x_n; \theta) = P(X_1 = x_1)...P(X_n = x_n)$

if the $X_i$ are discrete random variables, or

(7)  $L(x_1, ..., x_n; \theta) = f_{X_1}(x_1)...f_{X_n}(x_n)$

if the $X_i$ are continuously distributed random variables, where of course $f_{X_i}(x_i)$ is the probability density function of $X_i$.

(ii) Find the value(s) of $\theta$ such that $L(x_1, ..., x_n; \theta)$ is maximum.

If $L$ is a sufficiently well behaved function, then the calculus can be used to obtain the maximizing value(s) of $\theta$ as follows. Find the value(s) of $\theta$ such that

(8)  $\dfrac{dL}{d\theta} = 0,$

and then check to see which, if any, of these $\theta$'s yields the maximum value(s) of $L$.

Sometimes the maximum value(s) of $L$ can be found without using the calculus. This is especially true when the parameter $\theta$ can assume only a finite set of values, as the following example shows.

EXAMPLE 1  Suppose $X$ is a random variable that can take on the values 1, 2, ..., $m$ with equal probability so that

$P(X = k) = 1/m$

for $k = 1, 2, ..., m$. However, suppose we do not know the exact value of $m$. If we take a random sample $(x_1, x_2, ..., x_n)$ of $X$ values, we can form the likelihood of that sample:

$$L(x_1, x_2, ..., x_n; m) = \begin{cases} 0 & \text{if } m < \max\{x_1, x_2, ..., x_n\}, \\ (1/m)^n & \text{otherwise.} \end{cases}$$

This function takes on its maximum value when $m$ is smallest possible. Therefore the maximum likelihood estimate of $m$ is the largest value of $x_i$ in the sample.

EXAMPLE 2  Suppose we perform a number of coin-tossing experiments $\mathscr{E}_1$,

F

$\mathscr{E}_2$, ..., $\mathscr{E}_n$ where each $\mathscr{E}_i$ consists in tossing the same coin until the first head is obtained. Let $X_i = k$ where $k$ is the number of the trial (toss) upon which the first head is observed. Assume that there is some question concerning the fairness of the coin; we would like to estimate the probability $p$ of obtaining a head in an individual toss of the coin. It can be shown that

$$p(X_i = k) = q^{k-1}p,$$

which results from the solution of exercise 6 in section 3.1.

Since none of the experiments $\mathscr{E}_i$ has any effect on the others, the $X_i$ are independent and identically distributed. A sample $(x_1, ..., x_n)$ corresponding to the $n$-tuple $(X_1, ..., X_n)$ of random variables is thus random. Therefore, a maximum likelihood estimate of $p$ is possible:

$$
\begin{aligned}
L(x_1, ..., x_n; p) &= P(X_1 = x_1)...P(X_n = x_n) \\
&= (pq^{x_1-1})(pq^{x_2-1})...(pq^{x_n-1}) \\
&= p^n(1-p)^{\left(\sum_{i=1}^{n} x_i\right)-n}
\end{aligned}
$$

If we let $k' = \sum_{i=1}^{n} x_i$, then

$$L = p^n(1-p)^{k'-n}.$$

Clearly $L$ vanishes when $p = 0$ and $p = 1$, so the maximum lies between 0 and 1. To find it we form[3]

$$
\begin{aligned}
\frac{dL}{dP} &= np^{n-1}(1-p)^{k'-n} - p^n(k'-n)(1-p)^{k'-n-1} \\
&= p^{n-1}(1-p)^{k'-n-1}\{n(1-p)-(k'-n)p\} = 0,
\end{aligned}
$$

which results in the equation

$$n - np + (k'-n)p = 0$$

or

$$p = \frac{n}{k'} = n \left/ \sum_{i=1}^{n} x_i \right.$$

This can be shown to yield a maximum for $L$, and so

$$\hat{p} = n \left/ \sum_{i=1}^{n} x_i \right.$$

is the maximum likelihood estimate of $p$ given the sample $(x_1, ..., x_n)$.

---

3  Those without calculus can skip the technical details and proceed directly to the estimate.

EXAMPLE 3 Suppose we are sampling from a population which is known to be normally distributed with zero mean, so that if $(x_1, x_2, ..., x_n)$ is a random sample, each of the random variables $X_i$ has $N(x; 0, \sigma)$ as its distribution. Suppose also that $\sigma^2$ is not known. To obtain the maximum likelihood estimate of $\sigma^2$, first form the likelihood of obtaining $(x_1, x_2, ..., x_n)$

$$L(x_1, ..., x_n; \sigma^2) = \frac{1}{\sigma^n (2\pi)^{n/2}} e^{-\sum_{i=1}^{n} \frac{x_i^2}{2\sigma^2}}$$

and[4] then take the natural logarithm of $L$ to obtain

$$(9) \quad \ln L = -\frac{n}{2}(\ln \sigma^2 + \ln 2\pi) - \frac{1}{2\sigma^2} \sum_{i=1}^{n} x_i^2.$$

Setting the derivative of (9) equal to zero, we obtain

$$(10) \quad \hat{\sigma}^2 = \frac{1}{n} \sum_{i=1}^{n} x_i^2$$

as the maximum likelihood estimate of $\sigma^2$.

EXAMPLE 4 It can be shown that a random variable whose distribution has the density function

$$f_X(x) = \frac{a}{\pi} \frac{1}{a^2 + x^2} \quad (a > 0)$$

possesses no moments, i.e., $E(X^r)$ exists for no positive integer $r$ (in section 3.2 we showed this for $r = 1$). Thus it is not possible to estimate $a$ from a random sample $(x_1, x_2, ..., x_n)$ of values of $X$ by means of the method of moments. The method of maximum likelihood is, however, available. The likelihood function in this case is

$$(11) \quad L(x_1, x_2, ..., x_n; a) = \frac{a^n}{\pi^n} \frac{1}{(a^2 + x_1^2)...(a^2 + x_n^2)},$$

and if we form the natural logarithm of $L$, we obtain

$$\ln L = n \ln a - n \ln \pi - \sum_{i=1}^{n} \ln(a^2 + x_i^2),$$

and finally

$$(12) \quad \frac{d(\ln L)}{da} = \frac{n}{a} - \sum_{i=1}^{n} \frac{2a}{a^2 + x_i^2}.$$

---

4 Those without calculus can skip the technical details and proceed directly to the estimate.

Equating this derivative to zero, we obtain the best estimate for $a$. If $n = 1$, this estimate is

$$\hat{a} = \pm x_1.$$

Since $a$ takes only positive values, we can discard $-x_1$ as a possible estimate.

When more than one parameter is to be estimated, partial differentiation can be used to maximize the likelihood function. Thus in the case of a sample $(x_1, x_2, ..., x_n)$ from a normal population with unknown $\mu$ and $\sigma^2$, the maximum likelihood estimates of $\mu$ and $\sigma^2$ can be shown to be

$$\hat{\mu} = \bar{x} \quad \text{and} \quad \hat{\sigma}^2 = m_2 = \frac{1}{n} \sum_{i=1}^{n} (x_i - \bar{x})^2$$

using partial differentiation. Since we have already observed that $m_2$ is not an unbiased estimate of $\sigma^2$, a maximum likelihood estimate need not, in general, be unbiased. However, maximum likelihood estimates are consistent.

Maximum likelihood estimates of parameters can also be computed for samples that are not random. For more information about maximum likelihood and other methods of estimation see Chakravarti et al. (1967, pp. 278ff.) and Hodge and Lehmann (1964, chapter 8).

EXERCISES

1 In exercise 3 of section 4.2 you were asked to show that the method-of-moments estimate of the parameter $m$ was not altogether satisfactory. Can the method of maximum likelihood be used in this case to obtain a more satisfactory result? Remember that the likelihood function is the probability of obtaining the sample obtained, given the parameter to be estimated. (Hint: calculus is not necessary.)

2 Assume that the random variable $X$, equal to the number of typographical errors per page in a certain text, has the Poisson distribution so that

$$P(X = i) = \lambda^i e^{-\lambda}/i! \quad (i = 0, 1, 2, ...).$$

Suppose on $n$ pages selected at random there were $x_1, x_2, ..., x_n$ errors. Use[5] these values to obtain a maximum likelihood estimate of $\lambda$.

3 Suppose we have the same sampling situation as in example 2 of section 4.1, except that we wish to estimate $\sigma^2 = pq$.
   (a) Form the likelihood function for a random sample $(x_1, ..., x_n)$.
   (b) Using the calculus,[6] show that

$$\hat{\sigma}^2 = \left(\frac{1}{n} \sum_{i=1}^{n} x_i\right) \left(1 - \frac{1}{n} \sum_{i=1}^{n} x_i\right)$$

5 Those without calculus should, of course, omit this question.
6 Those without calculus should, of course, omit this part of the question.

is the maximum likelihood estimate of $\sigma^2$ given the random sample $(x_1, \ldots, x_n)$.

(c) How are the estimates $\hat{\sigma}^2$ above and $\hat{p}$ in example 2 of section 4.1 related?

4  Suppose a language containing a set of $N$ root morphemes at some time $t = 0$ replaces these roots by others with the same meaning in such a way that the probability of losing $n$ of the $N$ roots by time $t > 0$ is given by the expression

$$P_n(t) = \binom{N}{n} e^{-N\beta t}(e^{\beta t} - 1)^n.$$

Suppose further that at time $t_0$, $n_0$ of the original $N$ roots had disappeared from the language.

(a) Given the values $t_0$ and $n_0$ show that the likelihood function for estimating $\beta$ is

$$L(t_0, n_0; \beta) = \binom{N}{n_0} e^{-N\beta t_0}(e^{\beta t_0} - 1)^{n_0}.$$

(b)[7] Find the maximum likelihood estimate of $\beta$.

---

[7] Those without calculus should, of course, omit this part of the question.

# 5
# Hypothesis testing

## 5.1

A *statistical hypothesis* is a statement about the nature of the distribution of a random variable. The purpose of this section is to explain how such hypotheses can be tested.

In example 2 of section 2.4, we have already treated one such statistical hypothesis (due to A.S.C. Ross) and indicated how such a hypothesis might be tested. There we gave Ross's argument to the effect that the hypothesis that two languages, $L_1$ and $L_2$, say, are not genetically related is equivalent to the following hypothesis:

(H) *If, in a set of N Indo-European roots, $n_1$ have a cognate in $L_1$ and $n_2$ a cognate in $L_2$, then the distribution of the random variable R equal to the number of roots with cognates in both languages is governed by chance alone.*

In section 2.4 we showed that under this hypothesis $R$ has the hypergeometric distribution, that is

$$(1) \quad P(R = r) = \frac{\binom{n_2}{r} \binom{N-n_2}{n_1-r}}{\binom{N}{n_1}}.$$

To test this hypothesis, we can actually list the $N$ Indo-European roots, find $n_1$, $n_2$, and $r$, and finally compute $P(R \geq r)$, the probability of obtaining at least $r$ cognates by chance. If this probability is small, we would tend to reject the hypothesis H and assume that $L_1$ and $L_2$ are related. If $P(R \geq r)$ were not small, then it might be possible that this number of cognates could have occurred by chance, and no case could be made on probabilistic grounds for the two languages to be any more closely related than any other pair of Indo-European languages. This is a special case of Fisher's exact test (see section 5.3 and Bradley, 1968, pp. 195–203), as we remarked earlier, and is a very commonly used test.

A few observations about the above situation may prove helpful. Two kinds of error can be made by the researcher: (i) he may reject the hypothesis when it is in fact true, and (II) he may accept the hypothesis as being true when it is indeed false. Our main aim in this section is to control errors of the first kind.

In terms of our example, we may reject the hypothesis H that there is no genetic relation between $L_1$ and $L_2$ when there is, indeed, no such relation; this is an error of the *first kind* or a *type* I error. On the other hand, we may accept the hypothesis of no genetic relationship when it is false; this is an error of the *second kind* or a *type* II error. The other two possibilities – accepting H when it is true, and rejecting H when it is false – present no problem.

In general, the researcher can guard against type I errors by the methods that we shall outline below. However, type II errors are especially problematical to the linguistic researcher for two reasons. First, he has less control over them than he does over type I errors, and secondly he is often in the position where being able to accept the statistical hypothesis yields positive scientific results. Ross has guarded against this situation by forming his hypothesis in such a way that rejection yields the positive results, i.e., indicates a genetic relationship between the two languages. We shall return to this point shortly.

In the example considered above we tend to reject the hypothesis of chance occurrence if the number of shared cognate roots is high. If we were to choose a *critical value c* and reject the hypothesis whenever the observed value $r$ of $R$ is greater than or equal to $c$, then the probability of making a type I error is

(2) $\quad P(R \geqq c) = \sum_{i \geqq c} P(R = i)$

$\qquad\qquad$ = probability of obtaining a value of $R$ as high as $c$ under the given hypothesis.

This probability $P(R \geqq c) = \alpha$ of rejecting the hypothesis H when it is true can be controlled by first choosing $\alpha$, the *critical level*, and then finding $c$ so that

$\qquad P(R \geqq c) = \alpha.$

Generally $\alpha$ is chosen to be small when type I errors are especially hazardous; critical levels like 0.05, 0.01, and 0.005 are common, so that for example, if

(3) $\quad P(R \geqq c_{0.01}) = 0.01,$

then rejecting the hypothesis at $r \geqq c_{0.01}$ renders the probability of a type I error at 0.01. However, if the observed value $r$ of $R$ were $r < c_{0.01}$, this would not necessarily mean we could automatically assume H to be valid. Indeed, if

$\qquad P(R \geqq r) = 0.05,$

this would mean that the probability of obtaining an observed number of cognates at least as high as $r$ is quite small, 0.05, but not as small as the critical level, 0.01.

There is a distinct asymmetry to the situation just described. The hypothesis H is an assertion that there is nothing to the claim that $L_1$ and $L_2$ are related.

Thus in employing $c_{0.01}$ in expression (3), if we obtain an observed $r \geq c_{0.01}$ and reject the hypothesis, we are claiming $L_1$ and $L_2$ are related, with a probability of 0.01 that we are rejecting the hypothesis when it is true. On the other hand, if $r < c_{0.01}$, we are not in a position to accept H without incurring the possibility of an error of type II.

Hence in order to be able to make a certain claim C about the values of some random variable $X$, we try to arrange, as Ross did, to make the hypothesis H that there is nothing to C and then proceed to test H with an eye to rejecting it, and hence espousing C, if the value observed for $X$ is sufficiently improbable. For this reason H is often called a *null hypothesis* because, as indicated above, it usually amounts to a denial of some particular claim C. A null hypothesis H concerning a random variable $X$ is usually of the form: $X$ has some particular distribution $F_X$, the *null distribution*, from which such probabilities as

(4)  $P(X \geq c)$,

(5)  $P(|X| \geq c)$,

(6)  $P(|X - E(X)| \geq c)$,

can be computed. Then a critical or significance level is chosen and, depending on the needs of the problem, a critical value $c$ is obtained in (4), (5), (6), etc., so that the probability in question equals $\alpha$. $c$ is the *critical value* and $\alpha$ is the corresponding *level of significance* or *critical level*.

Thus a *hypothesis test* consists of making a null hypothesis about the distribution of $X$, choosing a level of significance $\alpha$, choosing a probability configuration like (4), (5), (6) or something else appropriate for a *critical region*, obtaining a critical value $c$ corresponding to $\alpha$ for that particular critical region, and rejecting or accepting the hypothesis according as the observed value of $X$ lies in the critical region or not. If the critical region is $X \geq c$ (as in (4)) or $X \leq c$,

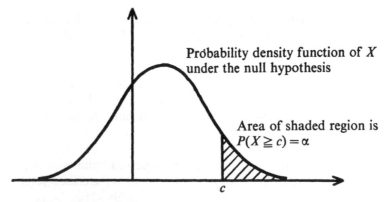

FIGURE 5.1  Upper tailed hypothesis test: if the area of the shaded region $P(X \geq c)$ equals $\alpha$, then reject the hypothesis at the $\alpha$-level when the observed value of $X$ is greater than or equal to $c$, i.e., when it falls in the critical region $X \geq c$.

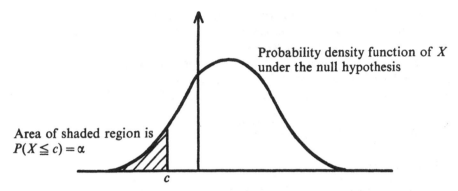

FIGURE 5.2   Lower tailed hypothesis test: reject the hypothesis if $X$ is less than or equal to $c$.

the test is called a *one-tailed test* as illustrated in figures 5.1 and 5.2. If, on the other hand, one wishes to use a critical region like $|X| \geq c$ or $|X - E(X)| \geq c$, the test is called a *two-tailed test*, as illustrated in figure 5.3. Whether a one-tailed or a two-tailed test is appropriate depends on the specific problem under consideration, as we shall have ample opportunity to see below.

EXAMPLE 1   Ross's scheme to test for significant relationships between languages is an example of an (upper) one-tailed hypothesis test. It is often complicated by the amount of computation necessary to obtain $P(R = r)$. For example, Ross (1950) cites, in his table 6, that in $N = 1860$ Indo-European roots there are $n_1 = 1184$ Italo-Celtic cognates and $n_2 = 1165$ Greek cognates, and of these $r = 783$ are common cognates in both languages. This means that to find $P(R \geq r)$ in this case we must compute, by (1) and (2),

$$P(R \geq 783) = \sum_{i \geq 783} \binom{1165}{i} \binom{1860 - 1165}{1184 - i} \bigg/ \binom{1860}{1184},$$

which is a prodigious task.

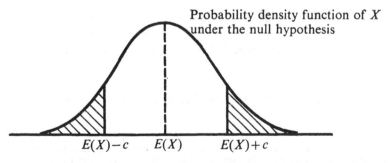

FIGURE 5.3   Two tailed hypothesis test: the critical region is $|X - E(X)| \geq c$. If the shaded area above it is $P(|X - E(X)| \geq c) = \alpha$, then we reject the null hypothesis at the $\alpha$-level, when the observed value of $X$ lies in the critical region. In this case there is a probability $\alpha$ of being wrong.

Since the hypergeometric distribution has been tabulated, in Lieberman and Owen (1961), for $N \leq 50$ and for certain selected values of $N$ greater than 50, a careful choice of $N$ may render Ross's test easier to use.

If $n_1$ and $n_2$ are large enough, as in the present example, then $R$ can be shown to have approximately the normal distribution $N(r, E(R), \sigma_R)$; see Lancaster (1969, p. 35). From exercise 7 in section 3.2 we find that

$$E(R) = \frac{n_1 n_2}{N} \quad \text{and} \quad \sigma_R{}^2 = \frac{n_1 n_2 (N-n_1)(N-n_2)}{N^2(N-1)}.$$

In the present case,

$$E(R) = 741.59 \quad \text{and} \quad \sigma_R{}^2 = 100.33.$$

Using a correction for continuity,[1]

$$P(R \geq 783) = 1 - N(783 - \tfrac{1}{2}, 741.59, \sqrt{100.33}).$$

By (20) in section 3.3, this can be written

$$P(R \geq 783) = 1 - N(4.08, 0, 1) = 0.00002$$

(which is highly indicative of a genetic relationship).

We shall return to this problem again in section 5.3 where other criteria for relationship between languages will be considered.

EXAMPLE 2   In Rebecca Hayden's study of American English (1950), there were 325 occurrences of /j/ in 65,122 phonemes. If we assume that Roberts's survey (1965) is over a sufficiently large corpus that his relative frequency 0.0036 of /j/ (see his p. 94) is a good approximation of the probability of /j/ and that both samples are random, then we can make the null hypothesis that both samples come from the same population for which the probability of selecting /j/ is $p = 0.0036$.

Hayden's sample can be taken as 65,122 Bernoulli trials with probability $p = 0.0036$ of obtaining /j/ in each trial. Thus, if the random variable $X_i$ ($i = 1, 2, \ldots, 65{,}122$) is defined to take the value 1 if the $i$th phoneme is /j/ and 0 otherwise, then with

$$\bar{X} = \sum_{i=1}^{65,122} X_i = \text{number of occurrences of /j/ in a 65,122-phoneme running text,}$$

---

1   In cases where discrete distributions are approximated by continuous ones it is useful to introduce this correction for continuity (sometimes called Yates's correction). In the particular case with which we are dealing it is not, strictly speaking, necessary because, to five decimal places, we obtain the same value for $P(R \geq 783)$ when the correction is not applied. For a normal approximation the correction is useful when the variance is small.

the null hypothesis can be stated as follows: $\overline{\overline{X}}$ has a binomial distribution with $n = 65,122$ and $p = 0.0036$, that is

$$P(\overline{\overline{X}} = k) = \binom{65,122}{k} (0.0036)^k (0.9964)^{65,122-k},$$

Since $np \geq 5$, the approximation of this binomial distribution by the normal distribution is adequate. Using this normal approximation, we can write

$$P(\overline{\overline{X}} \leq k) = N(k, np, \sqrt{npq}) = N(k, 234.4, \sqrt{233.6}),$$

which, according to (20) in section 3.3, can be written

$$P(\overline{\overline{X}} \leq k) = N\left(\frac{k-234.4}{\sqrt{233.6}}, 0, 1\right).$$

Let us perform a two-tailed test by choosing the critical value $c$ for the deviation of $\overline{\overline{X}}$ from $E(\overline{\overline{X}})$ corresponding to some significance level $\alpha$, i.e., let us choose $c$ such that

(7) $P(|\overline{\overline{X}} - E(\overline{\overline{X}})| \geq c) = \alpha.$

The expression

(8) $|\overline{\overline{X}} - E(\overline{\overline{X}})| \geq c$

is equivalent to the disjunction of

(9) $E(\overline{\overline{X}}) + c \leq \overline{\overline{X}}$

and

(10) $\overline{\overline{X}} \leq E(\overline{\overline{X}}) - c,$

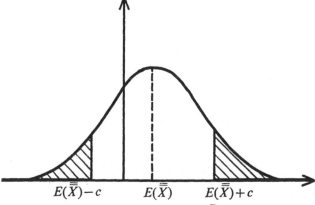

$$E(\overline{\overline{X}}) - c \qquad E(\overline{\overline{X}}) \qquad E(\overline{\overline{X}}) + c$$

FIGURE 5.4  The heavy lines to the right of $E(\overline{\overline{X}}) + c$ and to the left of $E(\overline{\overline{X}}) - c$ correspond to positions where $|\overline{\overline{X}} - E(\overline{\overline{X}})| \geq c$. The area of the shaded region is, of course, $P(|\overline{\overline{X}} - E(\overline{\overline{X}})| \geq c)$.

i.e., (8) holds if and only if either (9) or (10) holds (see figure 5.4). In the present case, the events $E(\bar{X}) + c \leq \bar{X}$ and $\bar{X} \leq E(\bar{X}) - c$ are disjoint events for $c > 0$, so

$$P(|\bar{X} - E(\bar{X})| \geq c) = P(\bar{X} - E(\bar{X}) \geq c) + P(\bar{X} - E(\bar{X}) \leq -c)$$

$$= 1 - P(\bar{X} < E(\bar{X}) + c) + P(\bar{X} \leq E(\bar{X}) - c)$$

$$= 1 - N(234.4 + c, 234.4, \sqrt{233.6})$$

$$+ N(234.4 - c, 234.4, \sqrt{233.6})$$

since $E(\bar{X}) = np = 234.4$ and $\sigma_{\bar{X}}^2 = 233.6$.

Using (20) in section 3.3, we can write $P(|\bar{X} - E(\bar{X})| \geq c)$ as follows:

$$P(|\bar{X} - E(\bar{X})| \geq c) = 1 - N\left(\frac{c}{\sqrt{233.6}}, 0, 1\right) + N\left(\frac{-c}{\sqrt{233.6}}, 0, 1\right)$$

$$= 1 - N\left(\frac{c}{15.28}, 0, 1\right) + N\left(\frac{-c}{15.28}, 0, 1\right).$$

The function

$$\frac{1}{\sqrt{2\pi}} e^{-t^2/2}$$

in the expression

$$N(x, 0, 1) = \frac{1}{\sqrt{2\pi}} \int_{-\infty}^{x} e^{-t^2/2} \, dt$$

is symmetric with respect to the vertical coordinate axis, and so

$$N(-x, 0, 1) = 1 - N(x, 0, 1).$$

Thus

$$(11) \quad P(|\bar{X} - E(\bar{X})| \geq c) = 2\left\{1 - N\left(\frac{c}{15.28}, 0, 1\right)\right\}.$$

If we set the level of significance at, say, 0.01, we must choose $c$ so that

$$P(|\bar{X} - E(\bar{X})| \geq c) = 0.01.$$

From a table of the normal distribution, we can obtain the result

$$N(2.58, 0, 1) = 0.995,$$

so that the right side of (11) equals 0.01. Thus if $c/15.28 = 2.58$ or

$$c = 15.28 \times 2.58 = 39.42$$

in expression (11), we have

$$P(|\bar{X} - 234.4| \geq 39.42) = 0.01.$$

Hayden's sample yields the value 325 for $\bar{X}$, and hence

$$\bar{X} - 234.4 = 90.6,$$

which is well beyond the critical value. Therefore in rejecting the hypothesis that $\bar{X}$ has binomial distribution with $p = 0.0036$, we have a probability of less than 0.01 of being wrong, i.e., of committing an error of type I. Thus there is over-whelming evidence that $\bar{X}$ is not binomially distributed with probability $p = 0.0036$. This is probably because, on the one hand, the $X_i$'s may not be independent (that is Hayden's sample may not be random) and, on the other, Hayden's sample is a very specialized one with perhaps a different value of $p$.

In general, when a null hypothesis takes the form *the random variable X has normal distribution with mean $\mu$ and variance $\sigma^2$*, then in terms of expression (6) *the equation relating the critical value c to the level of significance $\alpha$ is*

(12) $P(|X-E(X)| \geq c) = 2\{1-N(c/\sigma, 0, 1)\} = \alpha.$

EXAMPLE 3    Let us try the same procedure for the phoneme /ž/. In Hayden (1950), the author observed 19 instances of /ž/'s in 65,122 phonemes. Under the null hypothesis that $\bar{X}$, the number of occurrences of /ž/ in 65,122 trials, is binomially distributed with $p = 0.0003$, Roberts's relative frequency, we note that $\mu = np = 19.54$, $\sigma = \sqrt{npq} = \sqrt{19.53} = 4.42$, and

$$P(|\bar{X}-19.54| \geq c) = 2\{1-N(c/4.42, 0, 1)\}.$$

If we choose $\alpha = 0.2$ even, then for

$$N(c/4.42, 0, 1) = 0.90$$

we must have

$$c/4.42 = 1.28,$$

using a table of the normal distribution. Thus $c = 5.66$ and

$$P(|\bar{X}-19.54| \geq 5.66) = 0.2.$$

Hayden obtained a value of $\bar{X}$ equal to 19 so that $\bar{X}-19.54$ is very near zero and well below 5.66; there is therefore no need to reject the null hypothesis in this case because Hayden's result could well have occurred by chance.

For this example, Hayden obtained 19 for /ž/ in her sample, so

$$|\bar{X}-19.54| = 0.54.$$

From a table of the normal distribution we find that

$$P(|\bar{X}-19.54| \geq 0.54) = 2\{1-N(0.54/4.42, 0, 1)\} = 2\{1-N(0.12, 0, 1)\}$$
$$= 2(1-0.5478) = 0.90.$$

Thus the probability of obtaining a result with a deviation from 19.54 worse than Hayden's result under the null hypothesis is 0.90.

This apparent agreement with the null hypothesis may have arisen because the effect of the non-independence in Hayden's sampling method is small for /ž/.

The technique of examples 2 and 3 can be extended to obtain the *normal test for the difference of two percentages*. Suppose we have two random samples of size $N_1$ and $N_2$ in which $x_1$ and $x_2$ items respectively exhibit a certain attribute $A$. Under the null hypothesis that the two samples are from the same population, if both $N_1$ and $N_2$ are large enough, then the difference

$$d = \frac{x_1}{N_1} - \frac{x_2}{N_2}$$

can be shown to be approximately normally distributed with $\mu = 0$ and

(13) $\sigma^2 = pq(1/N_1 + 1/N_2)$.

Consequently,

(14) $P(|d| \geq c) = 2\{1 - N(c, 0, \sigma)\} = 2\{1 - N(c/\sigma, 0, 1)\}$.

Since $p$ is unknown, we must resort to one of the methods of estimation developed in chapter 4. Since the samples are assumed to be random and independent of one another, the likelihood of the two samples is

$$L(x_1, x_2) = \binom{N_1}{x_1} p^{x_1} q^{N_1 - x_1} \binom{N_2}{x_2} p^{x_2} q^{N_2 - x_2}.$$

Thus

$$\ln L = \ln \binom{N_1}{x_1} \binom{N_2}{x_2} + (x_1 + x_2) \ln p + [N_1 + N_2 - (x_1 + x_2)] \ln(1 - p),$$

and hence

$$\frac{d(\ln L)}{dp} = \frac{x_1 + x_2}{p} - \frac{N_1 + N_2}{1 - p} + \frac{x_1 + x_2}{1 - p}.$$

Setting this derivative equal to zero, we obtain the following estimate for $p$:

(15) $\hat{p} = \dfrac{x_1 + x_2}{N_1 + N_2}$,

and so the corresponding value of $\hat{\sigma}$ can be computed from (13). Empirical evidence indicates that good results are obtained when $N_1 \hat{p} > 5$ and $N_2 \hat{p} > 5$.

EXAMPLE 4   Hayden (1950, p. 220) obtained $x_1 = 24{,}378$ vowels in her sample of $N_1 = 65{,}122$ phonemes, and Roberts (1965, p. 112) obtained $x_2 = 20{,}240{,}466$

vowels in a sample of $N_2 = 56{,}056{,}231$ phonemes. Are the ratios $x_1/N_1$ and $x_2/N_2$ significantly different? To answer this question we use the test just described. The estimate of $p$ given by (15) is

$$\hat{p} = \frac{24378+20240466}{65122+56056231} = 0.3615,$$

and that of $\sigma$ is, by (13),

$$\hat{\sigma} = \sqrt{(0.3611)\,(0.6389)\,(0.0000154)} = \sqrt{3611}\,\sqrt{6389}\,\sqrt{1540} \times 10^{-7}$$

$$= 60.09 \times 79.93 \times 39.24 \times 10^{-7} = 1.855 \times 10^{-2} = 0.01855.$$

Since

$$\frac{x_1}{N_1} = \frac{24{,}378}{65{,}122} = 0.3743$$

and

$$\frac{x_2}{N_2} = \frac{20{,}240{,}466}{56{,}056{,}231} = 0.3611,$$

(16) $d = 0.0132$.

If we choose $\alpha = 0.01$, then by a table (Pearson and Hartley 1962) of the normal distribution

$$2\{1-N(c/\sigma, 0, 1)\} = 0.01$$

when $c/\sigma = 2.58$, or rather when $c = 0.05$. Thus, by (12),

(17) $P(|d| \geqq 0.05) = 0.01$.

From (17) it is clear that the difference $d$ of (16) is not significant at the 0.01 level. By (12) again

$$P(|d| \geqq 0.0132) = 2\{1-N(0.0132/\hat{\sigma}; 0, 1)\}$$

$$= 2\{1-N(0.69; 0, 1)\} = 0.4902.$$

Hence the vowel count yields no evidence that the population of Hayden's sample is different from that of Roberts's.

For more about the normal test for the difference of two percentages and other normal tests see Wadsworth and Bryan (1960, pp. 231–7).

EXERCISES

1 Use the normal test for the difference of two percentages to test the null hypothesis that Roberts's and Hayden's samples come from the same population in the case of

the following phonemes. Remember that Roberts's sample contains $N_1 = 56,056,231$ and Hayden's $N_2 = 65,122$ phoneme-tokens.

|  | $x_1$ (Roberts) | $x_2$ (Hayden) |
|---|---|---|
| (a) /n/ | 3,524,624 | 5,179 |
| (b) /d/ | 1,704,583 | 2,091 |
| (c) /g/ | 486,751 | 745 |
| (d) /θ/ | 235,305 | 286 |

2 A publishing house has a theoretical norm for its typesetters of an average of 9 typographical errors in 10 pages of set text. An applicant for the position of typesetter, in a trial run of 10 pages, made 12 typographical errors. Let the random variable $X$ be defined equal to the number of errors in a 10-page run. Then we can assume that $X$ follows the Poisson distribution with $\lambda = 9$, so

$$P(X = k) = 9^k e^{-9}/k!$$

Is the performance of the applicant significantly different from the norm at any of the levels 0.05, 0.01, 0.005?

3 The number $X$ of articles in a 50-word passage chosen at random from a certain novel is known to have the Poisson distribution with $\lambda = 3$. Test whether the following observations of articles in 50-word passages are significant at the 0.01, 0.05, or 0.005 levels of significance: (a) 4, (b) 5, (c) 6, (d) 7, (e) 10, (f) 12, (g) 13.

4 Suppose that, among $N = 15$ roots in a certain language family, two languages $L_1$ and $L_2$ of the family possessed $n_1 = 10$ and $n_2 = 9$ respectively of these roots and that they possessed $r = 7$ of these in common. What is the probability of their sharing 7 or more roots in common under the null hypothesis that they are unrelated?

## 5.2
### SOME GOODNESS-OF-FIT TESTS

#### 5.2.1 *The $\chi^2$-test for goodness-of-fit*

In the previous section we indicated a method for testing whether a single sample relative frequency of an event $E$ departs significantly from a hypothetical probability $p$ of $E$. Suppose now we wish to test simultaneously whether a number of sample relative frequencies of exhaustive mutually exclusive events $E_1, E_2, ..., E_m$ depart significantly from a corresponding hypothetical set of probabilities of these events. Suppose the hypothetical probabilities in question are

$$p_1, p_2, ..., p_m$$

and the corresponding observed relative frequencies in a random sample of size $n$ are

$$f_1/n, f_2/n, ..., f_m/n.$$

A commonly used measure of the deviation of the (observed) relative frequencies from the (expected) probabilities is

(1) $\quad Q = \sum_{i=1}^{m} \frac{(f_i - np_i)^2}{np_i}.$

If $Q$ is large, the deviation of the relative frequencies from the probabilities is large. It was shown by Karl Pearson in 1900 that if the sample in question is random and if the values $np_1, np_2, \ldots, np_m$ of the expected frequencies *are each not less than* 5, then $Q$ is well approximated by the $\chi^2$-distribution with $m-1$ 'degrees of freedom.' Since this distribution is extensively tabulated, $Q$ can be used to test for significant departures of the observed frequencies from the expected frequencies. The use of the $Q$ of expression (1) in conjunction with the $\chi^2$-distribution as indicated above constitutes the $\chi^2$-*test of goodness-of-fit*.

In general the procedure is as follows. Since it is reasonable to reject the null hypothesis that the relative frequencies represent a chance deviation from the hypothetical probabilities when $Q$ is large, we choose our critical value $c$ using the equation

$P(Q \geq c) = \alpha,$

and so the $\chi^2$-goodness-of-fit test is a one-tailed test. The values of the $c$'s for various degrees of freedom and various levels of significance $\alpha$ are tabulated in Pearson and Hartley (1962).

A word about *degrees of freedom* is in order here. Formally, the number of degrees of freedom of a statistic $S$, depending on a set of sample frequencies $(f_1, \ldots, f_n)$, is the maximum number of $f_i$'s that are not determined by the structure of $S$. Thus the number of degrees of freedom of $S$ is the number of independent facts embodied in $S$. For example, the statistic $Q$ in expression (1) depends on sample frequencies $f_1, f_2, \ldots, f_m$ whose sum is known to be $n$. The facts embodied in $Q$ are the frequencies $f_1, \ldots, f_m$. However, because their sum is known to be $n$, they are not independent. If we know all but one of them, say $f_m$, the value of $f_m$ is determined by the rest. Indeed,

$f_m = n - (f_1 + f_2 + \ldots + f_{m-1}),$

and so there are only $m-1$ independent facts embodied in the set of sample frequencies $f_1, \ldots, f_m$ known to add up to $n$. Thus $Q$ has only $m-1$ degrees of freedom. If information about the $f_i$'s were used to obtain the $p_i$'s (then the degrees of freedom of $Q$ might be decreased even more, as we shall see in example 2 below.

EXAMPLE 1  Roberts's table of vowel frequencies (1965, p. 99) indicates that the relative frequencies of the various vowels among all vowels are as given in table 5.1 (rounded off to four decimal places in all cases).

TABLE 5.1

| Vowel phoneme | Relative frequency |
|---|---|
| /ə/ | 0.3273 |
| /i/ | 0.2573 |
| /e/ | 0.1311 |
| /a/ | 0.1283 |
| /u/ | 0.0528 |
| /æ/ | 0.0427 |
| /o/ | 0.0426 |
| /ɔ/ | 0.0178 |

In his first decile (the tenth of his sample containing the most frequent words), Roberts (pp. 81–2) obtains the frequencies given in table 5.2.

One might ask the question: do the results of the first decile vary significantly from those of the whole sample? If we take the relative frequencies of the whole sample as the theoretical distribution, and make the null hypothesis that the frequencies of the first decile have this distribution, then if the $\chi^2 = Q$ is significant for some $\alpha$ with seven degrees of freedom (there are eight vowels), we can reject the hypothesis that the first decile has the same distribution as the whole survey, with a probability of $\alpha$ or less that we shall be rejecting a true hypothesis. In table 5.3 we have computed some of the values necessary for determining $Q$.

If we check a $\chi^2$-table, we find that with seven degrees of freedom

$$P(Q \geqq 25) = 0.001,$$

so, having computed only part of the sum in expression (1), we find that the departure of the first decile from the whole sample is highly significant and cannot be assigned to chance variation. This departure of the first decile may be explainable in part in terms of the following two factors: (i) We have assumed that the sample in each case is random which, as noted in exercise 6 of section

TABLE 5.2

| Vowel phoneme | Frequency |
|---|---|
| /ə/ | 5,685,407 |
| /i/ | 4,246,797 |
| /a/ | 2,294,815 |
| /e/ | 2,150,847 |
| /u/ | 939,488 |
| /æ/ | 700,525 |
| /o/ | 694,213 |
| /ɔ/ | 310,645 |
| TOTAL | 17,022,737 |

TABLE 5.3

| $v$ | $np_v$ | $f_v - np_v$ | $(f_v - np_v)^2/np_v$ |
|---|---|---|---|
| ə | 5 5 7 1 5 4 2 | 1 1 3 8 6 5 | 2 3 2 7 |
| i | 4 3 7 9 9 5 0 | | |
| e | 2 2 3 1 6 8 1 | | |
| a | 2 1 8 4 0 1 7 | | |
| u | 8 9 8 8 0 1 | 4 0 6 8 7 | 1 8 4 1 |
| æ | 7 2 6 8 7 1 | | |
| o | 7 2 5 1 6 9 | | |
| ɔ | 3 0 3 0 0 5 | | |
| TOTAL | 17 0 2 1 0 3 6 | | $Q > \quad$ 4 1 6 8 |
| | | | DF = 7 |

3.4, may be a questionable assumption. (ii) Although the first decile accounts for the bulk of the tokens (word instances) in Roberts's sample, its composition may be sufficiently different to account in part for the deviation encountered.

In some cases the expected frequencies $np_0$, $np_1$, ..., $np_m$ depend upon some parameter estimated from the data: for example, if the $p_i$'s follow the binomial distribution so that

$$p_i = \binom{m}{i} p^i q^{m-i}$$

for $i = 0, 1, 2, ..., m$ and the value of $p$ is to be estimated from the data using the method of moments so that

$$(2) \quad np = \bar{X} = \frac{1}{n} \sum_{i=0}^{m} i f_i$$

where $f_i$ is the number of sample values of $X$ equal to $i$, then

$$\hat{p} = \bar{X}/n.$$

In this case the frequencies must satisfy (2) as well, i.e. we must have

$$(3) \quad f_0 + f_1 + ... + f_m = n.$$

Thus, if all but two of the $m+1$ frequencies are known, then the remaining two can be found using equations (2) and (3). Thus in this case two degrees of freedom are lost, leaving $m-1$ independent choices of $f_i$ among the original $m+1$ frequencies.

Another situation of this kind arises when the expected relative frequencies follow the Poisson distribution.

EXAMPLE 2 In section 1.5, figure 1.7 gives the histogram for the number of articles in eighty 50-word passages selected at random from *Human Knowledge*

TABLE 5.4

| No. of articles | Observed frequency of passages with that no. of articles ($O_i$) | | Theoretical frequency using Poisson distribution ($E_i$) |
|---|---|---|---|
| 0 | 0 | | |
| 1 | 3 } 3 | | 5.00 |
| 2 | 10 | | 9.12 |
| 3 | 13 | | 13.61 |
| 4 | 20 | | 15.23 |
| 5 | 13 | | 13.63 |
| 6 | 9 | | 10.16 |
| 7 | 4 | | 6.50 |
| 8 | 4 | | |
| 9 | 3 } 8 | | 6.76 |
| 10 | 1 | | |

by Bertrand Russell. From this histogram we obtain the first two columns of table 5.4. As we observed in section 1.5, the sample mean is 4.475, and since the sample variance is around 4 we feel intuitively that the sample could have been drawn from a Poisson distribution with estimated mean $\hat{\lambda} = 4.475$ given to us by the method of moments. It does not seem unlikely that the occurrence of articles is one of those rare events that might be governed by a Poisson distribution. Using the null hypothesis that this Poisson distribution is the theoretical distribution of the random variable $X$ = number of articles in a 50-word passage, we lose two degrees of freedom, because the frequencies must add up to 80 and because the sample mean is fixed, requiring the frequencies to satisfy another equation. Thus only nine degrees of freedom remain.

The expected frequencies are given in the third column of table 5.4. The reader will note that the first two events, $X = 0$ and $X = 1$, and the last three events, $X = 8$, 9, and 10, were pooled to obtain the events $(0 \leq X \leq 1)$ and $(X \geq 8)$ in order that their expected frequencies be not less than 5. Expression (1) in this case yields

$$\chi^2 = \sum_{i=1}^{8} \frac{(f_i - 80p_i)^2}{80p_i} = \frac{(3-5)^2}{5} + \frac{(10-9.12)^2}{9.12} + \dots + \frac{(8-6.76)^2}{6.76} = 3.75.$$

Since we have pooled the information in the first two and the last three lines of the table, there remain only eight events $0 \leq X \leq 1$, $X = 2$, ..., $X = 7$, $X \geq 8$ after pooling. Since two degrees of freedom are lost already, there remain only six degrees of freedom. According to Lancaster (1969, p. 151) it is possible to argue that pooling of data does not result in the loss of a whole degree of freedom, but this effect is usually ignored. According to Pearson and Hartley (1962),

$$P(\chi^2 \geq 3.8) = 0.70372$$

with six degrees of freedom, so the probability of obtaining a deviation of $\chi^2$

greater than the one we obtained is greater than 0.7 under the null hypothesis that the sample was taken from a Poisson variate with $\lambda = 4.475$. Thus, under this null hypothesis we should obtain a larger $\chi^2$ in at least 7 samples out of 10. Although by no means absolute proof of the null hypothesis, this small value for $\chi^2$ is evidence for the truth of this hypothesis.

EXAMPLE 3   Fucks (1956) hypothesizes that in a text where the mean frequency of syllables/word is $\bar{\imath}$ the relative frequency $p_i$ of words with $i$ syllables is given by the formula

$$(4) \quad p_i = \frac{(\bar{\imath}-1)^{i-1}}{(i-1)!} e^{-(\bar{\imath}-1)}.$$

In analysing the first 501 words of 'La femme adultère,' one of the parts of *l'Exil et le Royaume* by Camus, R. Moreau (1964) obtained the results which appear in table 5.5.

In this problem there are four degrees of freedom because there are two restraints on the sample:

$$\sum_{i=1}^{6} f_i = 501 \quad \text{and} \quad \sum_{i=1}^{6} i f_i = 860,$$

where the first states that there are 501 words in the sample and the second that there are 860 syllables. Since both these equations are used to obtain $\bar{\imath}$, they constitute restrictions on the sample in the sense that if four of the $f_i$'s are known then the remaining two can be computed. Thus there are only four degrees of freedom. Although a pooling of information in the last three cells is indicated, let us first make the computation before pooling and then compare the result with that obtained after pooling. From table 7 of Pearson and Hartley (1962) with four degrees of freedom, we find that

$$P(Q \geq 4.60) = 0.33.$$

Thus $Q = 4.602$ is not significant at even the 0.1 level of significance and hence could easily have happened by chance.

Since $np_5$ and $np_6$ are less than 5, the $\chi^2$-approximation to $Q$ is unreliable.

TABLE 5.5

| No. of syllables per word | $p_i$ | $np_i$ | Observed frequencies $f_i$ | Difference | $(f_i-np_i)^2/np_i$ |
|---|---|---|---|---|---|
| 1 | 0.48843 | 244.702 | 251 | −6.298 | 0.1621 |
| 2 | 0.34999 | 175.345 | 160 | 15.345 | 1.3429 |
| 3 | 0.12539 | 62.824 | 74 | −11.176 | 1.9884 |
| 4 | 0.02995 | 15.006 | 14 | 1.006 | 0.0674 |
| 5 | 0.00537 | 2.688 | 1 | 1.688 | 0.0601 |
| 6 | 0.00077 | 0.385 | 1 | 0.615 | 0.9809 |

TABLE 5.5a

| No. of syllables per word | $p_4{}^*$ | $np_4{}^*$ | Observed frequency $f_4{}^*$ | Difference | $(f_4{}^* - np_4{}^*)^2/np_4{}^*$ |
|---|---|---|---|---|---|
| 4, 5, 6 | 0.03609 | 18.079 | 16 | 2.079 | 0.2391 |

However, if we pool the data of the last three rows in table 5.5 to obtain table 5.5a, we obtain $Q = 3.733$ with but two remaining degrees of freedom. In this case, table 7 of Pearson and Hartley (1962) yields

$$P(Q \geqq 3.8) = 0.15.$$

Thus $Q = 3.733$ is not significant at the 0.1 level. Although pooling the last three rows does not alter the situation much in regard to significance here, the two probabilities, 0.33 and 0.15, are quite different. In general the results when low-frequency cells are pooled tend to be more reliable.

In analysing a portion of 'Les muets,' another part of *l'Exil et le Royaume*, Moreau obtains the following results: number of words, 5035; number of syllables, 8253; $t = 1.639$; $Q = 147.491$. With four degrees of freedom, using $\alpha = 0.001$,

$$P(Q \geqq 18) = 0.001.$$

Therefore the actual frequencies depart significantly from those expected under the Fucks distribution given by (4).

Thus the results for 'La femme adultère' and for 'Les muets' seem to be at variance. We shall return to this point in the following subsection.

### 5.2.2 *The $\chi^2$-test for homogeneity*

The $\chi^2$-*test for homogeneity* can be used to test for heterogeneity among a number of different samples. Suppose $s$ independent random samples $j = 1, 2, ..., s$ are taken, of size $N_1, N_2, ..., N_s$, respectively. Divide the total range of sample values into $k$ mutually exclusive exhaustive categories ($i = 1, 2, ..., k$), the six syllable-per-word categories in example 3, for example. Let $O_{ij}$ be the number of values in sample $j$ which are observed to fall into category $i$. Under the null hypothesis that all the samples are random samples from the same population and hence have the same distribution, the expected number of values of sample $j$ observed to fall in category $i$ is given by the equation

$$E_{ij} = \left( \sum_{j=1}^{s} \frac{O_{ij}}{N} \right) N_j$$

where

$$N = \sum_{j=1}^{s} N_j.$$

TABLE 5.6

| $i$ | $j = 1$<br>'La femme adultère' | $j = 2$<br>'Les muets' | ROW TOTAL |
|---|---|---|---|
| 1 | 251 | 2848 | 3099 |
| 2 | 160 | 1335 | 1495 |
| 3 | 74 | 685 | 759 |
| $4 \leqq$ | 16 | 167 | 183 |
| COLUMN TOTAL | 501 | 5035 | 5536 |

It can then be shown (see Kendall and Stuart, 1966, p. 400) that

$$(5) \quad Q = \sum_{i=1}^{k} \sum_{j=1}^{s} \frac{(O_{ij} - E_{ij})^2}{E_{ij}} = \left( \sum_{i=1}^{k} \sum_{j=1}^{s} \frac{O_{ij}^2}{E_{ij}} \right) - N$$

has approximately the $\chi^2$-distribution with $(s-1)(k-1)$ degrees of freedom, again provided that the expected frequencies $E_{ij}$ are all greater than or equal to 5.

EXAMPLE 4  If we combine the information from 'La femme adultère' and 'Les muets,' we obtain table 5.6. Under the null hypotheses that the word lengths in syllables in the two chapters have identical distributions and that the two samples are independent, the expected number of 1-syllable words in 'La femme' and 'Les muets' should be

$$E_{11} = \frac{251 + 2848}{501 + 5035} (501) = 280.5$$

and

$$E_{12} = \frac{251 + 2848}{501 + 5035} (5035) = 2819,$$

respectively. The observed frequencies are, from table 5.6,

$$O_{11} = 251 \quad \text{and} \quad O_{12} = 2848.$$

The remaining expected frequencies are computed similarly:

$$E_{21} = 135.3, \quad E_{22} = 1360,$$

$$E_{31} = 68.69, \quad E_{32} = 690.3,$$

$$E_{41} = 16.53, \quad E_{42} = 166.2.$$

Thus

$$(6) \quad Q = \sum_{i=1}^{4} \sum_{j=1}^{2} \frac{(O_{ij} - E_{ij})^2}{E_{ij}} = 8.8.$$

Since there are four categories and two samples, the $\chi^2$-distribution that we are dealing with has $(3)(1) = 3$ degrees of freedom. From a table of the $\chi^2$-distribution with three degrees of freedom, we obtain

$$P(Q \geqq 8.8) = 0.03207,$$

significant at the 0.05 level but not at the 0.01 level. Thus we can make a mild case for the non-homogeneity of the combined sample, so Fucks's distribution might be valid for 'La femme adultère' while not holding for 'Les muets.' This question could be further elucidated by taking a larger sample from 'La femme adultère' to see if the close fit of the Fucks distribution was not an accident. In chapter 6, we return again to this problem of syllable counts.

Fucks's formula is a special case of the *translated Poisson distribution* given by

$$(7) \quad P(X = k) = \frac{(\lambda - n)^{k-n} e^{-(\lambda - n)}}{(k-n)!}$$

for $k = n, n+1, n+2, \ldots$ It is easy to show that

$$E(X) = \lambda \quad \text{and} \quad \sigma^2(X) = (\lambda - n)$$

when the distribution of $X$ is given by (7).

This distribution occasionally figures in other linguistic studies, as the following example indicates.

EXAMPLE 5   In *La Barraca* by Blasco Ibánez, it was found that in selecting fifty 50-word passages at random and noting the number $X$ of articles (*la, el, los, las, un, una*) the sample mean of $X$ was $\bar{X} = 6.860$ and the unbiased sample variance was 4.368 (see table 5.7 for the histogram of the sample values). In addition, no 50-word passage had fewer than two articles. The fact that there were never fewer

TABLE 5.7

| No. of articles | Passages with that no. of articles | Expected no. of passages using translated Poisson distribution |
|---|---|---|
| 2 | 0 ⎫ | |
| 3 | 1 ⎬ | 6.85 |
| 4 | 6 ⎭ | |
| 5 | 8 | 7.41 |
| 6 | 6 | 9.01 |
| 7 | 12 | 8.76 |
| 8 | 7 | 7.09 |
| 9 | 4 | 4.92 |
| 10 | 3 ⎫ | |
| 11 | 2 ⎬ | 5.96 |
| 12 | 1 ⎭ | |

than two articles per passage and that the sample variance $s_X^2$ was approximately two less than the sample mean $\bar{X}$ indicates a possible translated Poisson distribution for $X$ of the form

$$P(X = k) = \frac{(\lambda-2)^{k-2}e^{-(\lambda-2)}}{(k-2)!}$$

for $k = 2, 3, \ldots$ The method-of-moments estimate for $\lambda$ is $\hat{\lambda} = \bar{X} = 6.860$. If we form

$$Q = \frac{(7-6.85)^2}{6.85} + \frac{(8-7.41)^2}{7.41} + \frac{(6-9.01)^2}{9.01} + \ldots + \frac{(6-5.96)^2}{5.96},$$

we obtain $Q = 2.43$. Since there were seven categories whose probabilities must add up to 1 and we used up one degree of freedom in estimating the mean $\lambda = E(X)$, there are five degrees of freedom remaining. Table 7 in Pearson and Hartley (1962) yields the result

$$P(Q \geq 2.4) = 0.79147,$$

which means that, under the null hypothesis that $X$ has the translated Poisson distribution with $n = 2$, there is a probability of 0.79147 that the $Q$ computed from our sample will be greater than 2.4. Hence there is good evidence for the conjecture that $X$ follows the translated Poisson distribution with $n = 2$ and $\lambda = 6.860$.

### 5.2.3 The Kolmogorov test for goodness-of-fit

A test of goodness-of-fit often resorted to by psychologists and sociologists is the Kolmogorov test, or Kolmogorov-Smirnov test as it is sometimes called. Consider a random sample $(x_1, x_2, \ldots, x_n)$ of a random variable $X$ which is known to have a *continuous*[2] but unknown cumulative distribution function

$$P(X \leq x) = F_X(x).$$

The Kolmogorov test can be used to test the null hypothesis that $F_X(x)$ is equal to some particular cumulative distribution function $F_0(x)$. It consists in constructing the statistic

$$K = \sup_x |S_n(x) - F_0(x)|$$

---

2  This requirement ensures that the sample contains no tied observations, which tend to decrease the power of both the Kolmogorov test and the Smirnov test considered in exercise 8 at the end of section 5.2. Of course, even under the continuity assumption ties may occur as artefacts of imprecise measurement, for example, due to rounding off to a certain number of decimal places. However, these can be broken without effectively changing the power of the test. See Bradley (1968, chapter 3) for a discussion of this point.

where 'sup' is meant to signify the smallest real number larger than or equal to the values of $|S_n(x) - F_0(x)|$ for all $x$ (see section 0.2) and

$$S_n(x) = (1/n). \text{ (the number of indices } i \text{ such that } x_i \leqq x),$$

and then finding the probability of obtaining a value of $K$ larger than or equal to our observed value. A large value of $K$, of course, constitutes evidence for rejection of the null hypothesis that $F_X(x) = F_0(x)$. The approximate distribution of the statistic $K$ does not depend on the choice of $F_0(x)$. The percentage points of the distribution of $K$ have been tabulated by L.H. Miller (see Bradley, 1968, section 13.5 and table XIII) for values of $n$ between 1 and 100. In large samples, however,

$$P(d \geqq 1.36) = 0.05 \quad \text{and} \quad P(d \geqq 1.63) = 0.01$$

when $d = \sqrt{n} K$.

Since there is no mechanism for computing the effect upon the probabilities $P(K \geqq K_0)$ of using the sample to estimate parameters, we cannot legitimately use parameters estimated from the sample in our hypothesis of the $F_0(x)$. However, the sample could be divided into two parts, one of which might then be used to estimate the parameters in $F_0(x)$ and the other to construct $K$.

The following example shows another way of proceeding.

EXAMPLE 6   We have already noted in section 3.3 that the durations of sound bursts and silences in ordinary conversation are both approximately exponentially distributed (Cassotta et al. 1964). In a sample $(x_1, ..., x_{194})$ of 194 consecutive silences taken from a French recitation of *Visage de Napoléon*, the durations of silences were measured in $\mu$-(milli-)seconds. If $X$ represents the duration of a silence and the silences are assumed to be independent of one another, then under our hypothesis of an exponential distribution for $X$, the CDF of $X$ is

$$F_\beta(x) = F_X(x) = P(X \leqq x) = \frac{1}{\beta} \int_0^x e^{-t/\beta} \, dt = 1 - e^{-x/\beta}$$

for some choice of $\beta$.

Since the durations of silences are assumed to be continuously distributed, we can use the Kolmogorov test to test the hypothesis that $X$ is exponentially distributed. However, we must choose $\beta$. Owing to the fact that the sample of 194 silences was obtained by computer in the form exhibited in the first two columns of table 5.8, we cannot, as suggested above, separate the sample into two random samples and etsimate $\beta$ from one of them. The reader can verify for himself that $E(X) = \beta$, so the method-of-moments estimate of $\beta$ is

$$\bar{X} = \frac{1}{194} \sum_{i=1}^{194} x_i.$$

TABLE 5.8

| $x$ ($\mu$sec) | $S_{194}(x)$ | $F_{2000}(x)$ | $F_{1500}(x)$ |
|---|---|---|---|
| 30 | 0.010 | 0.015 | |
| 60 | 0.067 | 0.030 | |
| 90 | 0.113 | 0.044 | |
| 120 | 0.165 | 0.058 | |
| 150 | 0.201 | 0.072 | |
| 180 | 0.258 | 0.086 | |
| 210 | 0.273 | 0.100 | |
| 240 | 0.294 | 0.113 | |
| 270 | 0.330 | 0.126 | |
| 300 | 0.351 | 0.139 | |
| 330 | 0.356 | 0.152 | |
| 360 | 0.366 | 0.165 | |
| 390 | 0.387 | $\vdots$ | |
| 420 | 0.392 | | |
| 450 | 0.412 | | 0.259 |
| 480 | 0.423 | | |
| $\vdots$ | $\vdots$ | | |
| 3990 | 0.830 | | 0.955 |

However, since the exponential distribution is highly skewed, the sample values distant from the origin will have an undue effect on $\bar{X}$ and may tend to give too large an estimate of $\beta$. Thus, even if we could use the mean as an estimate in conjunction with the Kolmogorov test, it need not produce an optimal estimate.

If, however, we do compute $\bar{X}$, we find that it is approximately 2100 $\mu$sec. We can adopt the strategy of computing $F_\beta(x)$ for $\beta$'s near this value. Thus if we try $F_\beta(x)$ for $\beta = 2000$ $\mu$sec, we obtain the values in column 3 of table 5.8. We observe that

$$|S_{194}(330) - F_{2000}(330)| = |0.356 - 0.152| = 0.204,$$

so $\sup_x |S_{194}(x) - F_{2000}(x)| \geqq 0.204$ and hence

$$d = \sqrt{194}\,K \geqq \sqrt{194}\,(0.204) = 2.84.$$

However, $d = 1.63$ is significant at the 0.01, or, as we sometimes say, the 1 per cent level. Thus since in the present case $d \geqq 2.84$ it is highly significant. For $\beta > 2000$,

$$330/\beta < 330/2000,$$

and so

$$-330/2000 < -330/\beta.$$

Therefore

$$e^{-330/2000} < e^{-330/\beta},$$

and finally

$$F_\beta(330) = 1 - e^{-330/\beta} < 1 - e^{330/2000} = F_{2000}(330).$$

Thus for $\beta > 2000$

$$|S_{194}(330) - F_\beta(330)| > 0.204,$$

and so the corresponding $d$ is greater than 2.84 and hence significant. Trying a much smaller value, say $\beta = 1500$, we find that

$$|S_{194}(450) - F_{1500}(450)| = |0.412 - 0.259| = 0.153$$

and so

$$d = \sqrt{194}\, K \geqq \sqrt{194}\, (0.153) = 2.13,$$

which is again significant at the 0.01 level. We can reason again, as we did when $\beta = 2000$, that

$$|S_{194}(450) - F_\beta(450)| > 0.153$$

for any $\beta > 1500$, and hence reject the null hypothesis that the distribution of $X$ is exponential with $\beta \geqq 1500\ \mu$sec.

If we consider more of the data than we have given above, we find that

$$S_{194}(3990) = 0.830$$

and

$$F_{1500}(3990) = 1 - e^{-3990/1500} = 0.955.$$

Therefore

$$|S_{194}(3990) - F_{1500}(3990)| = F_{1500}(3990) - S_{194}(3990)$$
$$= 0.955 - 0.830 = 0.125.$$

Using the reasoning employed for $\beta > 2000$ we can show that for $\beta \leqq 1500$,

$$F_\beta(3990) \geqq F_{1500}(3990).$$

Therefore

$$|S_{194}(3990) - F_\beta(3990)| = F_\beta(3990) - S_{194}(3990) \geqq 0.125.$$

Therefore for $\beta \leqq 1500$

$$d = \sqrt{194}\, K \geqq \sqrt{194}\, (0.125) = 1.74,$$

and hence is significant at the 0.01 level.

What we have just shown in the last two paragraphs is that for $\beta \geqq 1500$, $d \geqq 2.13$ and for $\beta \leqq 1500$, $d \geqq 1.74$. Thus, for all possible values of $\beta$, we can

TABLE 5.9

|  | 0 | 1 |
|---|---|---|
| Observed frequency | $x$ | $16-x$ |
| Expected frequency | 8 | 8 |

reject (at the 0.01 level) the null hypothesis that the distribution of $X$ is exponential with paramater $\beta$.

It should be emphasized again that the Kolmogorov test is applicable only when the underlying hypothesized distribution is continuous. The following example illustrates this point very graphically.

EXAMPLE 7   Consider an experiment with only two outcomes 0 and 1. Under the null hypothesis that they are equally likely, $P(X = 0) = P(X = 1) = \frac{1}{2}$. Suppose that the experiment is replicated 16 times, each replication being independent of the others, and that $x$ of the trials result in 0. Thus $16-x = y$ result in 1, as in table 5.9. If we compute $K$ for the Kolmogorov test, we obtain

$$K = \sup \left\{ \left| \frac{x-8}{16} \right|, \left| \frac{y-8}{16} \right| \right\} = \sup \left\{ \frac{|x-8|}{16}, \frac{|x-8|}{16} \right\} = \frac{|x-8|}{16}.$$

The 0.05 level for $K$ is 0.327 (see table XIII in Bradley 1968). Thus, if the Kolmogorov test is applicable, $P(K \geq 0.327) = P(|x-8| \geq 5.24) = 0.05$.

Because the 16 replications of this experiment constitute a sequence of Bernoulli trials, we can obtain the exact probability that

(8)   $|x-8| \geq 5.24.$

Equation (8) holds for $x = 0, 1, 2, 14, 15, 16$. Because of the symmetry of the situation

$$P(|x-8| \geq 5.24) = 2 \left\{ \left( \frac{1}{2} \right)^{16} + 16 \left( \frac{1}{2} \right)^{16} + \frac{16 \times 15}{2} \left( \frac{1}{2} \right)^{16} \right\} = 0.004.$$

Thus the estimate of $P(|x-8| \geq 5.24)$ given by the Kolmogorov statistic is too large by a factor of 10. Hence, for example, if $K = 0.327$, the hypothesis that the deviation from the null hypothesis was significant would be rejected when in fact the actual probability $P(K \geq 0.327)$ would be near 0.004. Therefore much of the power of the Kolmogorov test is lost when the underlying distribution is discrete.

To see how the $\chi^2$-test stands up in this case, note that the $\chi^2$-test for goodness-of-fit yields

$$\chi^2 = \frac{(x-8)^2}{8} + \frac{\{(16-x)-8\}^2}{8} = \frac{(x-8)^2}{4}$$

with one degree of freedom. Since $P(\chi^2 \geqq 3.84) = 0.05$, the $\chi^2$-statistic yields the estimate

$$P\left(\frac{(x-8)^2}{4} \geqq 3.84\right) = 0.05.$$

In fact, the probability that

$$(x-8)^2/4 \geqq 3.84,$$

i.e., that

$$(x-8)^2 \geqq 15.36,$$

is equal to the sum of the probabilities that $x = 0, 1, 2, 3, 4, 12, 13, 14, 15, 16$, i.e.,

$$2\left\{\left(\frac{1}{2}\right)^{16} + 16\left(\frac{1}{2}\right)^{16} + \ldots + \frac{16!}{4!\,12!}\left(\frac{1}{2}\right)^{16}\right\} = 0.06.$$

For more information on the Kolmogorov test, its uses and limitations, see Bradley (1968, pp. 296ff.).

The Kolmogorov (or Kolmogorov-Smirnov) test, and the Smirnov test discussed in exercise 8 below, have received a certain amount of attention in the literature; see, for example, the papers by Rander Buch (pp. 76–9) and by J.J. Dreher and Elaine L. Young (pp. 156–69) in Doležel and Bailey (1969). In the latter paper (p. 162) the authors disregard the requirement of continuity on the underlying distribution when they compare (apparently) Mao Tse-tung's later Chinese-characters-per-segment distribution to that in his earlier work. If the inherent discreteness of the data has the effect of decreasing the power in the way we have indicated in example 7, then the apparent continuity of Mao's style may be only an artefact of their misuse of the test.

Rander Buch, on the other hand, may be on firmer ground, for the test tells him to reject the hypothesis of similarity of style and the test appears to give higher putative probabilities than actually exist, as was the case in example 7.

In the end, when there is the possibility of obtaining tied observations because of discontinuity in the underlying distribution of the variate to be tested, it is best to resort to the $\chi^2$-test with as many cells as possible. If, however, the null distribution of the variate under consideration is known to be continuous, the Kolmogorov and Smirnov tests appear to be superior to the $\chi^2$-test. See Bradley (1968, p. 300).

## EXERCISES

1 In Roberts (1965, pp. 116 and 117), we find the following information concerning the length of words in his sample according to syllables:

| No. of syllables | Frequency in total sample | Frequency considering only words in the first decile |
|---|---|---|
| 1 | 11,896,895 | 11,434,076 |
| 2 | 2,637,562 | 2,012,554 |
| 3 | 704,208 | 384,417 |
| 4 | 176,942 | 74,193 |
| 5 | 48,377 | 22,706 |
| 6 | 1,012 | 0 |
| 7 | 14 | 0 |
| 8 | 0 | 0 |

(a) Test the Fucks hypothesis (equation (4)) using the sample frequencies for the total population.

(b) Test the Fucks hypothesis using the first decile sample frequencies.

(c) Compute the relative frequencies $p_i$ ($i = 1, 2, ..., 8$) of $i$-syllable words in Roberts's total sample. Do the relative frequencies in the first decile depart significantly from those of the total sample?

2 (a) Suppose in a test for homogeneity that there are two samples ($j = 1, 2$) and two categories ($i = 1, 2$) and that the results are as given in table 5.10. Show that expression (5), in this case, reduces to

$$Q = N \frac{(ad - bc)^2}{(a+c)(b+d)(a+b)(c+d)}.$$

(b) The number of articles in 33 randomly selected 50-word narrative passages of *The Wapshot Chronicle* by John Cheever were counted with the following result:

| Number of articles | 0 | 1 | 2 | 3 | 4 | 5 | 6 | 7 | 8 | 9 | 10 |
|---|---|---|---|---|---|---|---|---|---|---|---|
| Number of passages | 1 | 2 | 1 | 6 | 8 | 2 | 3 | 5 | 3 | 1 | 1 |

In 33 randomly selected 50-word passages containing conversation in the same work we find:

| Number of articles | 0 | 1 | 2 | 3 | 4 | 5 | 6 | 7 | 8 |
|---|---|---|---|---|---|---|---|---|---|
| Number of passages | 2 | 5 | 10 | 7 | 4 | 3 | 1 | 0 | 1 |

Use the $Q$ in part (a) to test for homogeneity of the narrative sample and the conversation sample when category 1 is determined by the condition that the sample value is less than or equal to the median of the combined sample. Remember that the *median* of a sample is that sample value which divides the sample in two equal parts when all the

TABLE 5.10

|  | Sample 1 | Sample 2 |  |
|---|---|---|---|
| Category 1 | $O_{11} = a$ | $O_{12} = b$ |  |
| Category 2 | $O_{21} = c$ | $O_{22} = d$ |  |
|  | $N_1 = a+c$ | $N_2 = b+d$ | $N = a+b+c+d$ |

TABLE 5.11

| No. $X$ of personal pronouns | No. of passages with $X$ pronouns |
|---|---|
| 0 | 1 |
| 1 | 8 |
| 2 | 9 |
| 3 | 15 |
| 4 | 8 |
| 5 | 5 |
| 6 | 2 |
| 7 | 1 |
| 8 | 1 |

sample values are ranked in ascending order. This test can be applied to any two independent samples in order to test the null hypothesis that they are from the same population. The test can be found in many textbooks under the title *the median test. When N is small, the expected event-frequencies may be less than 5 and this test may be inaccurate.* In this case, the Fisher-Irwin test discussed in section 5.3 can be applied.

3 Find $E(X)$ for a random variable $X$ with the translated Poisson distribution given by expression (7). (Hint: in the summand of the expression

$$E(X) = \sum_{k=n}^{\infty} \frac{k(\lambda-n)^{k-n}}{(k-n)!} e^{-(\lambda-n)}$$

write the leftmost $k$ in the summand as $(k-n)+n$ and simplify.)

4 In *La Barraca* we found that in a random sample of fifty 50-word passages the histogram of the number of personal pronouns per passage was given by table 5.11. Find $\bar{X}$ and assuming that $X$ follows the Poisson distribution

$$P(X = k) = \frac{\lambda^k}{k!} e^{-\lambda}$$

estimate $\lambda$ and use the $\chi^2$-distribution to test the goodness-of-fit of the Poisson hypothesis.

TABLE 5.12

| No. of syllables | No. of words |
|---|---|
| 1 | 1441 |
| 2 | 400 |
| 3 | 119 |
| 4 | 28 |
| 5 | 10 |
| 6 | 1 |
| 7 | 1 |
| TOTAL | 2000 |

TABLE 5.13

| No. of syllables | Sample 1 | Sample 2 |
|---|---|---|
| 1 | 383 | 366 |
| 2 | 83 | 90 |
| 3 | 22 | 28 |
| 4 | 8 | 15 |
| 5 | 4 | 0 |
| 6 | 0 | 1 |
| TOTAL | 500 | 500 |

5 Use the $\chi^2$-distribution to test the goodness-of-fit of the Fucks distribution (given in expression (4)) to the data in table 5.12 obtained from 2000 consecutive words from *The Nigger of the Narcissus* by Joseph Conrad.

6 Two samples of 500 consecutive words each were taken from *Lord Jim* by Joseph Conrad and a syllable count made. The results are depicted in table 5.13. Use the $\chi^2$-distribution to test whether the null hypothesis that they were drawn from a homogeneous population can be rejected.

7 Two samples of 500 alternate consecutive words each were drawn from *Lord Jim*, i.e., every second word was drawn. A syllable count was made, the results of which appear in table 5.14. (a) Test the two samples for homogeneity. (b) Test the four samples from tables 5.13 and 5.14 for homogeneity. (c) What would you say about the dependence of the number of syllables in each of a consecutive pair of words upon one another? Is it marked or slight?

8 A variation of the Kolmogorov test, usually called the Smirnov test, can be used to test whether two random samples $(x_1, x_2, ..., x_n)$ and $(y_1, y_2, ..., y_m)$ are drawn from the same *continuous* distribution. Under the null hypothesis that they are drawn from the same distribution, the statistic

$$d = \max_x |S_n(x) - S_m(x)| \sqrt{(n+m)/nm}$$

where

$$S_n(x) = (1/n) \text{ (number of sample values } x_i \leqq x)$$

TABLE 5.14

| No. of syllables | Sample 1 | Sample 2 |
|---|---|---|
| 1 | 347 | 349 |
| 2 | 92 | 93 |
| 3 | 42 | 44 |
| 4 | 14 | 10 |
| 5 | 5 | 4 |
| 6 | 0 | 0 |
| TOTAL | 500 | 500 |

G

TABLE 5.15

| $x$ ($\mu$sec) | $S_{194}(x)$ | $S_{3003}(x)$ |
|---|---|---|
| 30 | 0.010 | 0.148 |
| 60 | 0.067 | 0.234 |
| 90 | 0.113 | 0.305 |
| 120 | 0.165 | 0.386 |
| 150 | 0.201 | 0.455 |
| 180 | 0.258 | 0.515 |
| 210 | 0.273 | 0.568 |
| 240 | 0.294 | 0.611 |
| 270 | 0.330 | 0.639 |
| 300 | 0.350 | 0.663 |
| 330 | 0.356 | 0.683 |

and

$$S_m(x) = (1/n) \text{ (number of } y_i \leqq x)$$

is distributed as the Kolmogorov statistic $d$ for values of $m$ and $n$ greater than 100.

The data in table 5.15 on length of silences was collected from 194 silences in the recitation of *Visage de Napoléon* and 3003 silences in a second French recitation. Use the Kolmogorov-Smirnov test to test the null hypothesis that the two samples came from the same distribution. For more on this test see Bradley (1968, pp. 288ff.).

## 5.3
### TESTS FOR INDEPENDENCE

### 5.3.1 *A $\chi^2$-test for independence*

Another important use of the $\chi^2$-distribution is to test the independence of two or more attributes. For simplicity we shall restrict ourselves to the case of two attributes which can be taken as the random variables $X$ and $Y$. To test for the independence of $X$ and $Y$, divide the values that $X$ can assume into $k$ mutually exclusive categories $\mathcal{X}_1, \mathcal{X}_2, ..., \mathcal{X}_k$ and the values that $Y$ can assume into $s$ mutually exclusive categories $\mathcal{Y}_1, \mathcal{Y}_2, ..., \mathcal{Y}_s$. Let $\alpha_i$ be the probability that $X$ takes on a value in category $\mathcal{X}_i$ and $\beta_j$ be the probability that $Y$ takes on a value in category $\mathcal{Y}_j$ where, of course,

$$\sum_{i=1}^{k} \alpha_i = \sum_{j=1}^{s} \beta_j = 1.$$

Let us choose a random sample of $N$ observations where $O_{ij}$ is the number of observations for which the random variable $X$ takes on a value in $\mathcal{X}_i$ and $Y$ a value in $\mathcal{Y}_j$. Then under the null hypothesis that $X$ and $Y$ are independent, the probability of selecting a sample observation for which $X$ takes on a value in $\mathcal{X}_i$ and $Y$ a value in $\mathcal{Y}_j$ is $\alpha_i\beta_j$, and so the expected number of our $N$ observations in $\mathcal{X}_i$ and $\mathcal{Y}_j$ simultaneously is $N\alpha_i\beta_j$. Then, as in the case of the goodness-of-fit test, we let

$$(1) \quad Q = \sum_{i=1}^{k} \sum_{j=1}^{s} \frac{(O_{ij}-N\alpha_i\beta_j)^2}{N\alpha_i\beta_j},$$

where large values of $Q$ indicate lack of independence. $Q$ follows the $\chi^2$-distribution, in this case with $ks-1$ degrees of freedom, if the $\alpha_i$'s and $\beta_j$'s are not estimated from the data. If, on the other hand, the parameters $\alpha_i$ and $\beta_j$ are estimated from the data:

$$\alpha_i = \frac{\text{frequency of X-values in category } \mathcal{X}_i}{N} = \frac{1}{N}\sum_{j=1}^{s} O_{ij}$$

$$\beta_j = \frac{\text{frequency of Y-values in category } \mathcal{Y}_j}{N} = \frac{1}{N}\sum_{i=1}^{k} O_{ij},$$

then there are only $(k-1)+(s-1)$ of these independent parameters, because

$$\sum_{i=1}^{k} \alpha_i = 1 \quad \text{and} \quad \sum_{j=1}^{s} \beta_j = 1,$$

and so the number of degrees of freedom is

$$ks-1-[(k-1)+(s-1)] = (k-1)(s-1).$$

EXAMPLE 1  In a random sample of fifty 50-word passages taken from *Riders in the Chariot*, we wish to test for a relationship between the number of nouns and the number of personal pronouns in each passage. In table 5.16 the results for a random sample of 50 passages are summarized. From the table it is clear that there is a tendency to use more nouns when fewer pronouns are used and vice versa. Thus there appears to be a dependence between nouns and pronouns. To see if this is significant we apply the $\chi^2$-test: if $X$ is the number of nouns and $Y$ the number of pronouns, then, under the independence assumption and using

TABLE 5.16

| No. of pronouns | No. of nouns | | | |
| --- | --- | --- | --- | --- |
| | 7 or fewer | 8 or 9 | 10 or more | ROW TOTAL |
| 3 or fewer | 4 | 3 | 8 | 15 |
| 4, 5, 6 | 3 | 9 | 6 | 18 |
| 7 or more | 9 | 5 | 3 | 17 |
| COLUMN TOTAL | 16 | 17 | 17 | 50 |

Each number in the nine interior cells of this table represents the number of passages falling into that cell category; thus, for example, four passages contained three or fewer pronouns and seven or fewer nouns.

the relative frequencies as estimates of the probabilities, we find that

$$\alpha_1 = 16/50, \qquad \alpha_2 = \alpha_3 = 17/50$$

and

$$\beta_1 = 15/50, \qquad \beta_2 = 18/50, \qquad \beta_3 = 17/50.$$

Thus

$$Q = \sum_{i=1}^{3} \sum_{j=1}^{3} \frac{(O_{ij} - 50\alpha_i\beta_j)^2}{50\alpha_i\beta_j}$$

$$= \frac{\left(4 - \dfrac{16 \times 15}{50}\right)^2}{\dfrac{16 \times 15}{50}} + \frac{\left(3 - \dfrac{17 \times 15}{50}\right)^2}{\dfrac{17 \times 15}{50}} + \cdots + \frac{\left(3 - \dfrac{17 \times 17}{50}\right)^2}{\dfrac{17 \times 17}{50}}$$

$$= 9.099$$

with four degrees of freedom. From a $\chi^2$-table we find that $P(Q \geq 9.0) = 0.06110$. Thus the results of applying this test appear to be inconclusive. Perhaps, if a larger sample were drawn, the dependence between nouns and pronouns might become more marked.

It is well to observe two things about the above example. First the expected value in the cell in the upper left-hand corner was

$$N\alpha_1\beta_1 = (50)\left(\frac{16}{50}\right)\left(\frac{15}{50}\right) = 4.8 \leq 5,$$

so the accuracy of $\chi^2$ might be impaired very slightly, and secondly we had to group the data in order to bring the expected frequency in each cell to near the required 5. A larger sample would bring the expected cell frequencies up and even allow for more differentiation and hence more cells. As a general rule, the accuracy of the $\chi^2$-test is increased by choosing the cells so that their respective expected frequencies are as near equal as possible.

### 5.3.2 $2 \times 2$ contingency tables

As a special case of the independence test, let us consider the $2 \times 2$ contingency tables. This concept arises naturally when we consider the problem of assessing the degree of connection between languages, first mentioned in example 2 of section 2.4. The technique to be discussed has been used (not altogether advisedly) by Chrétien and Kroeber (1937) and later by Chrétien alone (1943, 1956) to study linguistic data. In the first of these papers a history of the technique is given.

TABLE 5.17

|  | $L_1$ | $L_2$ | ... | $L_k$ |
|---|---|---|---|---|
| $f_1$ | − | − |  | + |
| $f_2$ | − | + |  | + |
| $\vdots$ |  |  |  |  |
| $f_N$ | − | − |  | − |

The problem put in a more general setting is as follows. Suppose we are given $k$ languages $L_1, L_2, ..., L_k$ and a set of $N$ (logically independent) features which each of the $k$ languages may or may not possess. In example 2 of section 2.4 the $k$ languages were all descended from Indo-European and the $N$ features could be characterized as possession or non-possession of a cognate to each of $N$ basic Indo-European roots. The general problem is to obtain a measure of the affinity or correlation between each of the

$$\binom{k}{2} = \frac{k(k-1)}{2}$$

pairs $\{L_i, L_j\}$ of languages under discussion.

We begin as in example 2 of section 2.4 by making an $N \times k$ table, each column corresponding to a language $L_n$ and each row to a feature $f_m$ (see table 5.17). If the $n$th language possesses feature $f_m$, then the $mn$th cell of the table contains $+$; if not, the cell contains $-$. Using such a table, we can construct the $2 \times 2$ table (table 5.18) for each pair $\{L_i, L_j\}$ of languages.

On looking back to example 2 of section 2.4, the reader will discover that in the notation of that example, we have table 5.19.

Using the reasoning of example 2 in section 2.4, we find that if $X$ is the random variable taking the value 'number of rows with a plus in both the $i$th and $j$th column,' then

TABLE 5.18

|  | No. of features possessed by $L_i$ | No. of features not possessed by $L_i$ | ROW TOTAL |
|---|---|---|---|
| No. of features possessed by $L_j'$ | $a$ | $b$ | $a + b$ |
| No. of features not possessed by $L_j'$ | $c$ | $d$ | $c + d$ |
| COLUMN TOTAL | $a + c$ | $b + d$ | $N$ |

TABLE 5.19

| $a = r$ | $b = n_j - r$ | $a+b = n_j$ |
|---|---|---|
| $c = n_i - r$ | $d = N - n_i - n_j + r$ | $c+d = N - n_j$ |
| $a+c = n_i$ | $b+d = N - n_i$ | $N$ |

$$(2) \quad P(X = a) = \binom{a+c}{a}\binom{b+d}{b} \bigg/ \binom{N}{a+b},$$

which corresponds to expression (10) in section 2.4.

In using (2) to test the hypothesis that $L_i$ and $L_j$ are unrelated, we are essentially applying an independence test. We are testing the hypothesis that, assuming there are $a+b$ properties possessed by $L_j$ and $a+c$ by $L_i$, the probability that a feature is possessed by both $L_i$ and $L_j$ is the product of the probability that it is possessed by $L_i$ times the probability that it is possessed by $L_j$.

Expression (2) can be used in other contexts to test the hypothesis that two dichotomies are independent of one another, as follows. If the properties that characterize the dichotomies are $A$ (a feature is possessed by $L_i$) and $B$ (a feature is possessed by $L_j$) and if in a sample of $N$ we have the results in table 5.20, then the distribution of the random variable $X = $ 'number of objects with both properties,' under the added hypothesis that the number of individuals having property $B$ is fixed at $a+b$ and the number of individuals with property $A$ is fixed at $a+c$, is given by (2). The null hypothesis that $A$ and $B$ are independent can be rejected at the $\alpha$ level if

$$P(X \leqq a) = \sum_{j \leqq a} P(X = j)$$

is less than $\alpha$ when $X$ can be assumed a priori to be small, and

$$P(X \geqq a) = \sum_{j \geqq a} P(X = j)$$

is less than $\alpha$ when $X$ can be assumed large. If no prior assumptions about $X$ can be made a two-tailed test should be used.

This test is well known as the Fisher-Irwin test, or Fisher's exact test. As an example, suppose in a series of $N = 8$ independent experiments we obtained $a+c = 6$ cases where $A$ held, $a+b = 4$ cases where $B$ held, and $a = 3$ cases

TABLE 5.20

|  | $A$ | Not $A$ | ROW SUM |
|---|---|---|---|
| $B$ | $a$ | $b$ | $a+b$ |
| Not $B$ | $c$ | $d$ | $c+d$ |
| COLUMN SUM | $a+c$ | $b+d$ | $a+b+c+d = N$ |

where both $A$ and $B$ held. $X$ can be greater than or equal to 3 in the following ways:

$$a = 3, \qquad b = 1, \qquad c = 3, \qquad d = 1,$$
$$a = 4, \qquad b = 0, \qquad c = 2, \qquad d = 2.$$

Therefore

$$P(X \geqq 3) = P(X = 3) + P(X = 4) = \binom{6}{3}\binom{2}{1} \bigg/ \binom{8}{4} + \binom{6}{4}\binom{2}{0} \bigg/ \binom{8}{4}$$

$$= \left( \frac{6 \times 5 \times 4 \times 2}{3!} + \frac{6 \times 5}{2!} \right) \frac{4!}{8 \times 7 \times 6 \times 5} = \frac{55}{70}.$$

Since $P(X = 3) = 40/70$, $X$ is likely to be equal to 3 more than half of the time.

As we have already remarked, this test is computationally unsatisfactory when $N$ is large. If $N$ is large, it has been observed, by Pearson and Hartley (1962, pp. 71–2), that the distribution of

(3) $\quad Q = N(r_{ij})^2$

is well approximated by the $\chi^2$-distribution with one degree of freedom, under the null hypothesis that $A$ and $B$ (or $L_i$ and $L_j$) are not related, when $r_{ij}$ takes the form[3]

(4) $\quad r_{ij} = \dfrac{ad - bc}{\sqrt{(a+b)\,(c+d)\,(a+c)\,(b+d)}}.$

(See also exercise 2 in section 5.2.) It can easily be shown that $r_{ij}$ has the following properties:

$-1 \leqq r_{ij} \leqq 1,$

if $b = c = 0$, then $r_{ij} = 1$,

if $a = d = 0$, then $r_{ij} = -1$.

In the language example, if $b = c = 0$, then $L_i$ possesses a particular feature if and only if $L_j$ possesses the same feature. Thus the two languages cannot be distinguished by the features under discussion, and hence must be highly related. On the other hand, if $a = d = 0$, then $L_i$ and $L_j$ are related 'negatively' in the sense that, if $L_i$ possesses a feature, then $L_j$ is sure not to have that feature. Finally, $r_{ij}$ would tend to be near zero if there were no particular tendency for $L_i$ and $L_j$ to share or eschew features simultaneously.

Because of the properties just listed Chrétien and Kroeber (1937) proposed $r_{ij}$ as a measure of the affinity of languages $L_i$ and $L_j$. In this paper the results

---

3   Sometimes a correction for continuity is used in (4), which results in a more complicated $r_{ij}$ (for a discussion of this point see Pearson and Hartley 1962, p. 72).

TABLE 5.21

|  | Greek | Non-Greek | TOTAL |
|---|---|---|---|
| Italo-Celtic | 783 | 401 | 1184 |
| Non-Italo-Celtic | 382 | 294 | 676 |
| TOTAL | 1165 | 695 | 1860 |

must be viewed with some scepticism because the features under consideration are not logically independent.

EXAMPLE 2　As an example of the technique outlined above, let us consider again the results of table 6 in Ross (1950). He found with $N = 1860$ Indo-European roots that $n_1 = 1184$ had cognates in Italo-Celtic ($L_1$), $n_2 = 1165$ had cognates in Greek ($L_2$), and $r = 783$ had common cognates in both languages. The counterpart of table 5.18 in this case is table 5.21. Then $r_{12}$ can be computed using expression (3):

$$r_{12} = \frac{(783)\,(294) - (401)\,(382)}{\sqrt{(1184)\,(676)\,(1165)\,(695)}} = 0.0957.$$

To see if this value is large enough to indicate a genetic relation between Italo-Celtic and Greek which could not have arisen by chance, let us test it for significance. By (3),

$$Q = N(r_{12})^2 = (1860)\,(0.0957)^2 = 17.03.$$

Since $Q$ has the $\chi^2$-distribution with one degree of freedom, we obtain from a $\chi^2$-table that

$$P(Q \geqq 10.83) = 0.001.$$

Thus $Q = 17.03$ is highly significant, the probability that such a value could occur by chance being distinctly less than 0.001. Compare this result with that obtained in example 1 of section 5.1. The two methods yield comparable results, in this case. In exercise 8 you are asked to check another case.

EXAMPLE 3　In a list of 74 features Chrétien and Kroeber (1937, p. 18) found that for Baltic and Italic $a = 7, b = 21, c = 22, d = 24$, for Baltic and Slavic, $a = 23$, $b = 5$, $c = 4$, $d = 42$, and for Baltic and Sanskrit $a = 10$, $b = 18$, $c = 17$, $d = 29$. Using expression (4) they obtained

$$r_{\text{B.I.}} = -0.23, \qquad r_{\text{B.Sl.}} = 0.74, \qquad r_{\text{B.Sa.}} = -0.01,$$

and using expression (3) they obtained

$$Q_{\text{B.I.}} = 3.914, \qquad Q_{\text{B.Sl.}} = 40.51 \qquad Q_{\text{B.Sa.}} = 0.0074.$$

Since

$$P(Q \geqq 0.016) = 0.9,$$
$$P(Q \geqq 3.84) = 0.05,$$
$$P(Q \geqq 10.83) = 0.001,$$

it follows that the probability that

$$Q_{\text{B.Sa.}} \geqq 0.0074$$

by chance is over 0.9, so the test provides no evidence for a significant relationship between Baltic and Sanskrit. On the other hand the probability that

$$Q_{\text{B.Sl.}} \geqq 40.51$$

by chance is well under 0.001, and so the test indicates strongly the existence of a causal relationship between Baltic and Slavic.

The case of Baltic and Italic is somewhat puzzling, for the probability of

$$Q_{\text{B.I.}} \geqq 3.914$$

is less than 0.05 and so it is significant at the 0.05 level. Thus the probability that this negative correlation of $-0.23$ could occur by chance is $\leqq 0.05$. This would seem to provide evidence for the existence of a causal relation which tends to produce, for a given feature, a minus sign for Baltic when Italic has a plus and vice versa.

More startling in this regard is the relationship that Chrétien and Kroeber found between Italic and Iranian. There $a = 4$, $b = 25$, $c = 27$, $d = 18$, which yield

$$r_{\text{It.Ir.}} = -0.46 \quad \text{and} \quad Q_{\text{It.Ir.}} = 16.03$$

where

$$P(Q_{\text{It.Ir.}} \geqq 16.03) < 0.0001.$$

This is apparently strong evidence that there is some causal mechanism which produces this negative correlation. However, such a mechanism is highly counter-intuitive. The seeds of this strange state of affairs are likely engendered by the methods of data collection and classification and the choice of features. Another possibility is that $r$ is an inappropriate measure of similarity between genetically related languages. This point is discussed in detail in Ellegård (1959).

EXERCISES

1 A few hypothetical examples may help to put certain of the results of this section in clearer perspective. Suppose that eight languages $L_1, L_2, ..., L_8$ are daughter languages of a language $L_0$ containing three roots and that, in this case, table 5.17 takes the following form:

| Roots | $L_0$ | $L_1$ | $L_2$ | $L_3$ | $L_4$ | $L_5$ | $L_6$ | $L_7$ | $L_8$ |
|---|---|---|---|---|---|---|---|---|---|
| 1 | + | − | + | + | + | − | − | − | + |
| 2 | + | + | − | + | − | + | − | − | + |
| 3 | + | + | + | − | − | − | + | − | + |

(a) Find $r_{01}$, $r_{03}$, $r_{06}$, $r_{07}$, and $r_{08}$.

(b) Find $r_{12}$ and $r_{45}$.

(c) Find $r_{14}$, $r_{78}$, $r_{26}$, $r_{23}$, and $r_{58}$.

(d) Find all possible $2 \times 2$ tables (like table 5.18) for which $N = a+b+c+d = 3$, and then for each of these tables find if possible a pair $(L_i, L_j)$ of languages from $L_0$, ..., $L_8$ to which this table corresponds. For example, the table

3  0
0  0

is the $2 \times 2$ table corresponding to $(L_0, L_8)$, and

2  0
0  1

is the $2 \times 2$ table corresponding to $(L_3, L_3)$.

(e) Suppose a new text of $L_0$ is discovered which contains a fourth root. The cognates of this new root give rise to the following new row in the above table of pluses and minuses:

| Roots | $L_0$ | $L_1$ | $L_2$ | $L_3$ | $L_4$ | $L_5$ | $L_6$ | $L_7$ | $L_8$ |
|---|---|---|---|---|---|---|---|---|---|
| 4 | + | − | − | − | + | + | + | + | − |

Find the new values of $r_{78}$, $r_{14}$, $r_{26}$, $r_{23}$, and $r_{58}$. Compare these values to those obtained in part (c).

(f) Given the results of (a)–(e) above, how do you feel about the appropriateness of $r_{ij}$ as a measure of linguistic affinity?

2 For general non-negative values of $a$, $b$, $c$, $d$ with $a+b+c+d = N$, if

$$r(a, b, c, d) = \frac{ad - bc}{\sqrt{(a+b)(c+d)(a+c)(b+d)}},$$

then show that (a) $r(a, b, c, d) = r(d, b, c, a)$, (b) $r(a, b, c, d) = r(a, c, b, d)$, and (c) $r(a, b, c, d) = -r(b, a, d, c)$.

Suppose that for one pair of languages $(L_A, L_B)$ $a = 10$, $b = c = 45$, $d = 100$ and for another pair $(L_C, L_D)$ $a = 100$, $b = c = 45$, $d = 10$. Assuming all these languages are daughter languages of some language $L_0$, make up a linguistic rationale to cover these two situations and compute $r_{AB}$ and $r_{CD}$.

3 If there are no borrowings of roots from one language into another, then knowing that $L_2$ is a daughter language of $L_1$ and that $L_3$ is a daughter language of $L_2$, what relationships do you feel should obtain among the values of $a$, $b$, $c$, and $d$ in the $2 \times 2$ tables for $(L_1, L_2)$, $(L_2, L_3)$, and $(L_1, L_3)$? Illustrate your answer using examples from the eight hypothetical languages in exercise 1.

4 (a) Show that

$$\lim_{d\to\infty} r(a, b, c, d) = \lim_{d\to\infty} \frac{ad-bc}{\sqrt{(a+b)(c+d)(a+c)(b+d)}} = \frac{a}{\sqrt{(a+b)(a+c)}}.$$

Ellegård (1959) observes that for data involving root correspondences

$$\rho = \frac{a}{\sqrt{(a+b)(a+c)}}$$

is more intuitively satisfying as a measure of the affinity between two genetically related languages. One can reason roughly as follows. If the daughter languages $L_1$ and $L_2$ of language $L_0$ are sufficiently distant from $L_0$ in time, then the roots of $L_0$ preserved in either daughter language will be a very small fraction of the total roots of $L_0$. Thus if the number of roots in $L_0$ is $N$, then the root cognates in $L_1$ and $L_2$ are given by the following $2 \times 2$ table:

|  | In $L_2$ | Not in $L_2$ |  |
|---|---|---|---|
| In $L_1$ | $a$ | $b$ | $a+b$ |
| Not in $L_1$ | $c$ | $d$ | $c+d$ |
|  | $a+c$ | $b+d$ | $N$ |

where $d$ is very much larger than $a+b+c$. If $L_0$ is a protolanguage like Indo-European, then, if it existed, it was bound to have a vast bank of roots not attested to in its recent descendents. Therefore it is reasonable to assume $\rho$ as a measure of affinity between $L_1$ and $L_2$.

(b) Show that $0 \le \rho \le 1$, and if $\rho = \rho(a, b, c)$, then $\rho(a, 0, 0) = 1$ and $\rho(0, b, c) = 0$.

(c) Interpret $\rho(a, 0, 0)$ linguistically.

(d) Interpret $\rho(0, b, c)$ linguistically.

(e) What happens to the function $A = N r(a, b, c, d)$ as $d \to \infty$? Can $\chi^2$ still be used to test the significance of $Q$ for very large (unknown) values of $d$? Why?

5 In a sample of 170 fifty-word passages drawn at random from the novel *Lord Jim* by Joseph Conrad, the following pairs, written in the form (number of nouns in the passage, number of pronouns), were obtained: (9, 3), (12, 2), (9, 5), (6, 10), (4, 5), (7, 6), (11, 1), (11, 2), (12, 5), (8, 6), (10, 8), (15, 4), (11, 4), (7, 1), (11, 2), (10, 4), (8, 10) (9, 5), (13, 2), (14, 2), (5, 10), (8, 6), (12, 7), (12, 7), (6, 11), (7, 8), (7, 7), (12, 2), (10, 8), (6, 8), (11, 7), (15, 1), (11, 6), (13, 4), (7, 8), (10, 6), (11, 2), (10, 4), (7, 5), (8, 6), (11, 6), (8, 11), (10, 6), (5, 10), (10, 3), (2, 5), (13, 2), (10, 7), (9, 6), (7, 7), (6, 9), (6, 6), (1, 12), (6, 6), (9, 8), (12, 10), (7, 9), (7, 9), (3, 16), (7, 6), (4, 16), (6, 11), (5, 11), (5, 10), (10, 6), (4, 15), (5, 9), (8, 10), (3, 12), (8, 11), (7, 7), (2, 14), (12, 6), (6, 9), (5, 9), (9, 8), (6, 12), (9, 9), (5, 12), (5, 9), (8, 11), (6, 16), (11, 6), (8, 8), (4, 12), (12, 0), (10, 6), (8, 2), (11, 0), (7, 6), (13, 1), (11, 3), (14, 1), (13, 1), (12, 2), (14, 1), (9, 2), (5, 4), (13, 0), (14, 1), (14, 1), (14, 1), (15, 1), (10, 6), (4, 7), (10, 3), (15, 0), (11, 3), (12, 5), (13, 2), (12, 4), (13, 2), (8, 2), (10, 5), (8, 8), (12, 5), (9, 6), (9, 5), (16, 0), (10, 4), (14, 1), (4, 10), (6, 8), (16, 6), (8, 6), (7, 11), (5, 6), (6, 12), (10, 7), (8, 9), (4, 10), (8, 4), (11, 8), (5, 6),

(15, 3), (11, 3), (8, 6), (9, 5), (6, 6), (7, 11), (11, 3), (9, 4), (10, 10), (14, 2), (6, 8), (8, 8), (13, 4), (10, 6), (9, 2), (6, 9), (10, 4), (14, 2), (13, 6), (9, 6), (11, 0), (12, 5), (6, 4), (8, 5), (1, 12), (11, 5), (9, 5), (6, 6), (6, 4), (8, 6), (8, 6), (12, 1), (9, 4), (9, 4), (5, 9), (11, 7). If $X$ is the number of nouns in a passage and $Y$ the number of pronouns in the passage, test this sample for the independence of $X$ and $Y$. Try to choose your cells so that the expected number of sample values per cell, in each case, lies between 8 and 17. The more equal these expected values are the more reliable is the $\chi^2$-test.

6 Suppose the following two independent samples of six observations each are drawn from a single random variable:

$$S_1 = (1, 2, 3, 4, 1, 1), \qquad S_2 = (1, 2, 6, 5, 8, 9).$$

(a) Find the median of the combined sample.

(b) Let category 1, as discussed in exercise 2(b) of section 5.2, be defined by the property of being less than or equal to the combined median of part (a) and set up the appropriate $2 \times 2$ table. Then test the sample for independence.

7 In order to compare the normal approximation to the hypergeometric distribution mentioned in section 5.1 and the $\chi^2$-approximation, use the former on the data of example 1 to ascertain the significance of the results obtained there. How would you check the results of example 3 for accuracy?

8 Ross observes $n_1 = 1184$ Celto-Italic cognates and $n_2 = 442$ Armenian cognates to the 1860 Indo-European roots considered with $R = 333$ agreements. Compute $r_{12}$ and $Q$ in this case. Use the methods of example 1 of section 5.1 to obtain $P(R \geq 333)$.

## 5.4
### COMPARISON OF MEANS OF NORMAL VARIATES

In this section we consider two new sampling distributions: the $t$-distribution introduced by W.S. Gosset in 1908 in a paper written under the pseudonym 'Student,' and the $F$-distribution introduced by R.A. Fisher. Both the $t$- and $F$-distributions are exact when the parent population from which the sample is drawn is normal, but the $t$-distribution is approximate otherwise. Empirical studies have shown that the $t$-distribution is not very sensitive to moderate departures from normality. The $F$-distribution, on the other hand, is sensitive to departures from normality, so it is best to use the $F$-distribution only for large samples, where there is some justification for normality assumptions due to the central limit theorem.

### 5.4.1 *The t-test of a single mean*

If $(x_1, x_2, ..., x_n)$ is a sample of a random variable $X$ which is known to have a normal distribution but with unknown mean and variance, then we can estimate the distribution mean $E(X)$ and variance $\sigma_X{}^2$ using the method of moments as

(1) $\quad \bar{X} = \dfrac{1}{n} \sum\limits_{i=1}^{n} x_i$

and

(2) $\quad s_X{}^2 = \dfrac{1}{n-1} \sum\limits_{i=1}^{n} (x_i - \bar{X})^2,$

respectively. However, there is no guarantee that these values are near the actual mean $\mu_X$ and variance $\sigma_X{}^2$ of the distribution of $X$. In case we are interested in null hypotheses about the distribution mean, for example the hypothesis that the mean is $\mu_0$, and have no interest in the value of $\sigma_X{}^2$, we can use (1) and (2) to compute

(3) $\quad t = \dfrac{\bar{X} - \mu_0}{\sqrt{s_X{}^2/n}}$

*which is known to have the t-distribution with $n-1$ degrees of freedom.*

This test can be used to locate the population mean in the following sense. If we calculate $\bar{X}$ and $s_X{}^2$ from a sample of size $n$, and we can obtain a value of $t$, say $t_{n-1}^{\alpha}$ corresponding to a critical value $\alpha$, from a table (e.g., table 9 of Pearson and Hartley 1962) of the $t$-distribution such that $P(|t| > t_{n-1}^{\alpha}) = \alpha$, then

$$P\left( \left| \dfrac{\bar{X} - \mu_0}{\sqrt{s_X{}^2/n}} \right| > t_{n-1}^{\alpha} \right) = \alpha.$$

Since the sum of the probability of a set and that of its complement is one, $P(|t| > t_{n-1}^{\alpha}) = \alpha$ if and only if

$$P(|t| \leq t_{n-1}^{\alpha}) = 1 - \alpha,$$

i.e.,

(4) $\quad P\left( \left| \dfrac{\bar{X} - \mu_0}{\sqrt{s_X{}^2/n}} \right| \leq t_{n-1}^{\alpha} \right) = 1 - \alpha.$

Thus with probability $1 - \alpha$ we have

$$-t_{n-1}^{\alpha} \leq \dfrac{\bar{X} - \mu_0}{\sqrt{s_X{}^2/n}} \leq t_{n-1}^{\alpha},$$

and so with the same probability we have

(5) $\quad \bar{X} - t_{n-1}^{\alpha} s_X/\sqrt{n} \leq \mu_0 \leq \bar{X} + t_{n-1}^{\alpha} s_X/\sqrt{n}.$

For example, if $\bar{X} = 5$, $s_X = 1$, and $n = 25$, then for $\alpha = 0.01$, say, a table of the $t$-distribution yields

$$t_{24}^{0.01} = 2.797,$$

TABLE 5.22

| No. of nouns $X$ | 5 | 6 | 7 | 8 | 9 | 10 | 11 | 12 | 13 | 14 | 15 | 16 |
|---|---|---|---|---|---|---|---|---|---|---|---|---|
| No. $f_X$ of passages containing $X$ nouns | 1 | 0 | 2 | 5 | 5 | 10 | 13 | 12 | 7 | 4 | 0 | 1 |

as the 0.01 critical value for 24 degrees of freedom, and (5) yields

$$5 - \frac{2.80}{5} \leqq \mu_0 \leqq 5 + \frac{2.80}{5}$$

or

$$4.44 \leqq \mu_0 \leqq 5.56.$$

Thus, in 99 cases in 100, $\mu_0$ will fall in the above range.

Expression (5) is sometimes said to give the $1 - \alpha$ *confidence limits* of the population mean $\mu_0$.

EXAMPLE 1  In sampling 50-word passages from *La Barraca* by Blasco Ibánez, we obtained the data in table 5.22 when counting $X$, the number of common nouns, in each passage. Since the *t*-test is relatively insensitive to departures from normality and since the sample is relatively large, the *t*-test should give reliable results. Using the figures in table 5.22, we obtain $\bar{X} = 10.90$, $s_X^2 = 4.13$, and $n = 60$. Thus with $\alpha = 0.01$, we find that $t_{59}^{0.01} = 2.66$ and so by expression (5)

$$10.90 - \frac{2.66\sqrt{4.13}}{\sqrt{60}} \leqq \mu_X \leqq 10.90 + \frac{2.66\sqrt{4.13}}{\sqrt{60}},$$

whence

$$10.20 \leqq \mu_X \leqq 11.60$$

with probability 0.99.

### 5.4.2 *The t-test of the difference of the means of two independent samples*

If $(x_1, x_2, ..., x_n)$ and $(y_1, y_2, ..., y_m)$ are two independent samples of random variables $X$ and $Y$ respectively, then, under the null hypothesis that (i) both $X$ and $Y$ are normally distributed with equal variance, i.e., $\sigma_X^2 = \sigma_Y^2 = \sigma^2$, and (ii) $\mu_X - \mu_Y = \delta$ both hold, an unbiased estimate of the common variance $\sigma^2$ is given by

$$(6) \quad s^2 = \frac{(n-1)s_X^2 + (m-1)s_Y^2}{n+m-2},$$

and the random variable

TABLE 5.23

| No. $i$ of articles in passage | No. $f_i$ of passages containing $i$ articles in | |
| :---: | :---: | :---: |
| | Narrative passages | Conversation passages |
| 0 | 1 | 2 |
| 1 | 2 | 4 |
| 2 | 4 | 10 |
| 3 | 7 | 7 |
| 4 | 12 | 4 |
| 5 | 10 | 3 |
| 6 | 9 | 1 |
| 7 | 10 | 0 |
| 8 | 5 | 1 |
| 9 | 2 | 0 |
| 10 | 2 | 0 |
| TOTAL | 64 | 32 |

$$(7) \quad t = (\bar{X} - \bar{Y}) - \delta \Big/ s \sqrt{\frac{1}{n} + \frac{1}{m}}$$

*has the t-distribution with $n + m - 2$ degrees of freedom.*

EXAMPLE 2   In sampling 50-word passages from a novel, we find good evidence that $X$, the number of articles in a 50-word passage which includes conversation, and $Y$, the number of articles in a 50-word passage not including any conversation, both have approximately the Poisson distribution. It is well known that for a Poisson variate $Z$ with mean $\lambda$, the variance also equals $\lambda$. Thus under some circumstances we might be able to assume that the $t$-test would yield approximately reliable results, but since we do not necessarily assume $\mu_X = \mu_Y$, we cannot assume $\sigma_X{}^2 = \sigma_Y{}^2$. However, a well-known *variance-stabilizing transformation*, the square root (Lancaster 1969, section 37.11), tends to stabilize the variance of the transformed Poisson variate $\sqrt{Z}$ at near $\frac{1}{4}$ and renders the distribution of $\sqrt{Z}$ approximately normal. We consider therefore

$$W = \sqrt{Z}$$

instead of $Z$. Table 5.23 summarizes the results of a random sample of sixty-four 50-word narrative-passages and a random sample of thirty-two 50-word conversation-passages drawn from *The Wapshot Chronicle* by John Cheever. If $X$ is the number of articles in a narrative passage and $Y$ the number in a conversation passage, then

$$\sqrt{\bar{X}} = \frac{1}{64} \sum_{i=0}^{10} \sqrt{i} f_i = 2.20,$$

$$s_{\sqrt{X}} = \sqrt{\frac{1}{63} \sum_{i=0}^{10} (\sqrt{i} - \sqrt{\bar{X}})^2} = 0.5616,$$

$$\sqrt{\bar{Y}} = \frac{1}{32} \sum_{i=0}^{10} \sqrt{i} f_i = 1.57,$$

and

$$s_{\sqrt{Y}} = \sqrt{\frac{1}{31} \sum_{i=0}^{10} (\sqrt{i} - \sqrt{\bar{Y}})^2} = 0.5979.$$

Then under the null hypothesis that $\mu_{\sqrt{X}} = \mu_{\sqrt{Y}}$,

$$s^2 = \frac{63s_{\sqrt{X}}^2 + 31s_{\sqrt{Y}}^2}{94} = 0.3293$$

and

$$t = \frac{(2.20 - 1.57) - 0}{s\sqrt{\frac{1}{32} + \frac{1}{64}}} = 5.11$$

with 94 degrees of freedom.

Since a table (e.g., table 9 in Pearson and Hartley 1962) of the $t$-distribution yields the result

$$P(|t| \geq 3.46) = 0.001$$

for 60 degrees of freedom and

$$P(|t| \geq 3.37) = 0.001$$

for 120 degrees of freedom, the fact that for a fixed probability, in this case 0.001, $t$ is a decreasing function of its degrees of freedom, yields the result

$$P(|t| \geq \beta) = 0.001$$

for some $\beta$ between 3.46 and 3.37 with 94 degrees of freedom. Our value of $t$, i.e., $t = 5.11$, clearly has a probability of less than 0.001 of being that large. Therefore we can reject the hypothesis that $\mu_{\sqrt{X}} = \mu_{\sqrt{Y}}$, and so conclude that the use of articles in passages containing conversation differs from that in narrative passages, at least in *The Wapshot Chronicle*.

One can proceed to find the confidence limits of $\delta$, say at the 0.999-level. As we have seen,

$$t = \frac{(2.20 - 1.57) - \delta}{s\sqrt{\dfrac{1}{32} + \dfrac{1}{64}}}$$

has the $t$-distribution with 94 degrees of freedom under the null hypothesis that the two means differ by $\delta$, and we know that

$$P(|t| \geq 3.46) = 0.001.$$

Therefore the probability that

$$\left| \frac{(2.20 - 1.57) - \delta}{s\sqrt{\dfrac{1}{32} + \dfrac{1}{64}}} \right| \geq 3.46$$

is less than 0.001, and so by an algebraic manipulation we obtain the result that the probability that

$$2.20 - 1.57 - 3.46\, s\sqrt{\frac{3}{64}} < \delta < 2.20 - 1.57 + 3.64\, s\sqrt{\frac{3}{64}}$$

and hence that

$$0.20 < \delta < 1.06$$

is greater than 0.999 under the null hypothesis.

### 5.4.3 Comparison of two sample variances

If $(x_1, x_2, ..., x_n)$ and $(y_1, y_2, ..., y_m)$ constitute random samples of a variate $X$ and a variate $Y$ respectively, both having a normal distribution, then under the null hypothesis that $\sigma_Y^2 = \sigma_X^2$ the quotient

(8) $\quad F = s_X^2 / s_Y^2$

*follows the F-distribution with $n-1$, $m-1$ degrees of freedom.*

The following example shows how this remark can be used to test for the equality of two variances.

EXAMPLE 3   In example 2 we tested the hypothesis that $\mu_{\sqrt{X}} = \mu_{\sqrt{Y}}$ under the assumption that $\sigma_{\sqrt{X}}^2 = \sigma_{\sqrt{Y}}^2$. Now we are in a position to check this latter assumption. From the information in example 2 we find that

$$F_{63,31} = \frac{(0.5616)^2}{(0.5979)^2} = \frac{0.3154}{0.3575} = 0.88$$

with 63 and 31 degrees of freedom. Since we could have chosen the names $X$ and $Y$ of the sample variates in example 2 in reverse order with equal likelihood (and so there is no reason to consider that one of the variances $\sigma_{\sqrt{X}}{}^2$ or $\sigma_{\sqrt{Y}}{}^2$ is larger than the other), we are testing the null hypothesis $\sigma_{\sqrt{X}}{}^2 = \sigma_{\sqrt{Y}}{}^2$ against the dual possibility that either $\sigma_{\sqrt{X}}{}^2 < \sigma_{\sqrt{Y}}{}^2$ or $\sigma_{\sqrt{Y}}{}^2 < \sigma_{\sqrt{X}}{}^2$. Thus we must use a two-tailed test. In this case it takes the form of finding the probability

$$P(F_{63,31} \leq 0.88 \text{ or } F_{31,63} \geq (0.88)^{-1}) = 2P(F_{31,63} \geq (0.88)^{-1}).$$

Since most tables of the $F$-distribution concern themselves only with the upper tail, it is best to consider that variance ratio $s_{\sqrt{Y}}{}^2/s_{\sqrt{X}}{}^2 = F_{31,63}$ which is $\geq 1$. Consulting a table – e.g. table 18 in Pearson and Hartley (1962) – we find that with 30 and 60 degrees of freedom

$$P(F \geq 1.22) = 0.25$$

and with 40 and 120 degrees of freedom

$$P(F \geq 1.18) = 0.25,$$

and that finally, since $F_{\nu_1\nu_2}^{\alpha}$ in $P(F \geq F_{\nu_1\nu_2}^{\alpha})$ is a decreasing function of $\nu_1$ and $\nu_2$,

$$F_{40,120}^{0.25} = 1.18 \leq F_{31,63}^{0.25} \leq F_{30,60}^{0.25} = 1.22.$$

Thus, since $(0.88)^{-1} = 1.14$,

$$P(s_{\sqrt{Y}}{}^2/s_{\sqrt{X}}{}^2 \geq 1.14) = P(s_{\sqrt{X}}{}^2/s_{\sqrt{Y}}{}^2 \leq 0.88) \geq 0.25$$

under the null hypothesis. Therefore

$$P(s_{\sqrt{X}}{}^2/s_{\sqrt{Y}}{}^2 \leq 0.88 \text{ or } s_{\sqrt{X}}{}^2/s_{\sqrt{Y}}{}^2 \geq 1.14) \geq 0.50,$$

and so we have no reason to reject the hypothesis $\sigma_{\sqrt{X}}{}^2 = \sigma_{\sqrt{Y}}{}^2$. Hence this tends to increase our confidence in the results of example 2.

### 5.4.4 An F-test for the equality of three or more means

If we wish to compare the means of three or more samples taken from normal populations with equal variance, one might think it appropriate to use the $t$-test to test the significance of the difference between any two. There is a danger in this procedure which can be illustrated as follows. If we are just testing the difference between the means of variates $X$ and $Y$ with $n_x + n_y - 2$ degrees of freedom, then for

$$t_{X-Y} = \frac{\bar{X} - \bar{Y}}{\sqrt{\dfrac{(n_x - 1)s_X{}^2 + (n_y - 1)s_Y{}^2}{n_x + n_y - 2}} \sqrt{\dfrac{1}{n_X} + \dfrac{1}{n_Y}}}$$

$P(t_{X-Y} \geq t_{n_x+n_y-2}^{0.05}) = 0.05$. Thus the probability of obtaining a significant $t_{X-Y}$ under the null hypothesis is 0.05. Suppose, on the other hand, that, rather than ust two, there are six samples of variates $X_1, X_2, ..., X_6$ which are independent, normally distributed, and have equal variance. Under these circumstances, if

$$t_{X_i - X_j} = \frac{\bar{X}_i - \bar{X}_j}{\sqrt{\dfrac{(n_{X_i} - 1)s_{X_i}^2 + (n_{X_j} - 1)s_{X_j}^2}{n_{X_i} + n_{X_j} - 2}} \sqrt{\dfrac{1}{n_{X_i}} + \dfrac{1}{n_{X_j}}}},$$

then surely $t_{X_1 - X_2}$, $t_{X_3 - X_4}$, and $t_{X_5 - X_6}$ are independent. Suppose we consider the null hypotheses:

$$H_1: \quad \mu_{X_1} = \mu_{X_2},$$

$$H_2: \quad \mu_{X_3} = \mu_{X_4},$$

$$H_3: \quad \mu_{X_5} = \mu_{X_6}.$$

The test of these hypotheses can be thought of as a sequence of three Bernoulli trials where the probability of a success, i.e., a value of $t_{X_i - X_j} \geq t_{n_i + n_j - 2}^{0.05}$, in each trial is 0.05. Thus the probability of obtaining one significant value in the three trials under the conjunction of the three null hypotheses is not 0.05, but according to the binomial distribution

$$\binom{3}{1}(0.05)(0.95)^2 = 0.135.$$

The probability of obtaining no significant value in the three trials is not 0.95, as is the case when one hypothesis is involved, but rather

$$(0.95)^3 = 0.86.$$

If other hypotheses are also included, e.g.,

$$H_4: \quad \mu_{X_1} = \mu_{X_3}$$

even though independence among the various $t_{X_i - X_j}$ may no longer be valid, the probabilities of various combinations of significance become even more complex.

These problems can be avoided by applying the following test based on the $F$-distribution. Consider $r$ samples $(x_{1k}, x_{2k}, ..., x_{n_k k})$ of the $r$ variates $X_k$ where for $k = 1, 2, ..., r$ each $X_k$ is assumed normally distributed with a common variance

$$\sigma_{X_k}^2 = \sigma^2.$$

The sample variance *within samples* is given by the expression

$$(9) \quad s_W^2 = \frac{1}{N - r} \sum_{k=1}^{r} \sum_{i=1}^{n_k} (x_{ik} - \bar{X}_k)^2,$$

and the sample variance *between samples* is given by

$$(10) \quad s_B^2 = \frac{1}{r-1} \sum_{k=1}^{r} n_k(\bar{X}_k - \bar{X})^2$$

where

$$N = \sum_{k=1}^{r} n_k, \quad \bar{X}_k = \frac{1}{n_k} \sum_{i=1}^{n_k} x_{ik}, \quad \bar{X} = \frac{1}{N} \sum_{k=1}^{r} \sum_{i=1}^{n_k} x_{ik}.$$

Sometimes it is useful to break down the computation of these sample variances using the following intermediate variables:

$$N = \sum_{k=1}^{r} n_k, \quad T_k = \sum_{i=1}^{n_k} x_{ik}, \quad T = \sum_{k=1}^{r} T_k, \quad S_k = \sum_{i=1}^{n_k} x_{ik}^2.$$

Then we can easily show that

$$(11) \quad s_W^2 = \frac{1}{N-r} \sum_{k=1}^{r} \left( S_k - \frac{T_k^2}{n_k} \right),$$

$$(12) \quad s_B^2 = \frac{1}{r-1} \left[ \left( \sum_{k=1}^{r} \frac{T_k^2}{n_k} \right) - \frac{T^2}{N} \right].$$

*Under the null hypothesis*

$$H: \quad \mu_{X_1} = \mu_{X_2} = \dots = \mu_{X_r},$$

*$s_W^2$ and $s_B^2$ are both estimates of the common variance $\sigma^2$, and so*

$$(13) \quad F = s_B^2 / s_W^2$$

*has the F-distribution with $r-1$ degrees of freedom associated with the numerator and $N-r$ associated with the denominator.*

If the differences between the sample means are significant, then $s_B^2$ should be significantly larger than $s_W^2$, and so a violation of H should result in a value of $F$ significantly larger than 1. Thus this test compares the null hypothesis with a simple alternative of $F$ significantly large, and hence is a one-tailed $F$-test, unlike the situation in example 3 above.

If in a particular situation $F$ turns out to be small enough so that $1/F$ is significantly large, then it is wise to check the assumptions to see if there is anything wrong with the original modelling of the situation.

EXAMPLE 4    In example 3 we saw how a variance-stabilizing transformation may be applied to a Poisson random variable. In 15 random samples of 50-word passages from the 15 sources listed in table 5.24, we considered the 15 variates

$X_k$ = number of articles in a passage drawn from source $k$.

TABLE 5.24

| $k$ | Name of $k$th source | No. $n_k$ in $k$th sample | Mean of $\sqrt{X_k}$, $\sqrt{\bar{\bar{X}}_k}$ | Standard deviation of $\sqrt{X_k}$, $s_{\sqrt{x_k}}$ |
|---|---|---|---|---|
| 1 | *W.C.* (conversation) | 32 | 1.57 | 0.5979 |
| 2 | *L.J.* (conversation) | 35 | 1.58 | 0.5182 |
| 3 | *S.M.* | 100 | 1.64 | 0.6044 |
| 4 | *A.B.R.* | 50 | 1.75 | 0.5635 |
| 5 | *Riders* | 50 | 1.91 | 0.4380 |
| 6 | *Vic.* | 50 | 1.92 | 0.6346 |
| 7 | *Nightmares* | 50 | 1.97 | 0.6108 |
| 8 | *L.J.* | 100 | 2.01 | 0.6351 |
| 9 | *H.K.* | 80 | 2.06 | 0.4823 |
| 10 | *Power* | 50 | 2.13 | 0.4016 |
| 11 | *Obs.* | 70 | 2.13 | 0.4302 |
| 12 | *G.M.* | 100 | 2.17 | 0.4608 |
| 13 | *W.C.* (narrative) | 64 | 2.21 | 0.5616 |
| 14 | *T.P.H.* | 67 | 2.31 | 0.3339 |
| 15 | *L.J.* (narrative) | 35 | 2.38 | 0.4981 |

The abbreviations have the following meaning: *W.C.* = *The Wapshot Chronicle* by John Cheever (two sources were used, passages containing some conversation and pure narrative passages, i.e., passages without conversation); *L.J.* = *Lord Jim* by Joseph Conrad (here there are three sources: (i) passages containing conversation, (ii) pure narrative, and (iii) a random sample of 50-word passages drawn without regard for conversation); *S.M.* = *The Solid Mandala* by Patrick White; *A.B.R.* = *The Autobiography of Bertrand Russell*; *Riders* = *Riders in the Chariot* by Patrick White; *Vic.* = *Victory* by Joseph Conrad; *Nightmares* = *Nightmares of Eminent Persons* by Bertrand Russell; *H.K.* = *Human Knowledge* by Bertrand Russell; *Power* = *Power: A New Social Analysis* by Bertrand Russell; *Obs.* = random sample from *The Observer* of January 1970; *G.M.* = random sample from the *Toronto Globe and Mail* of 22 January 1970; *T.P.H.* = *Theory and Processes of History* by F.J. Teggart. Except for those specially mentioned, all samples were chosen at random without regard to conversation.

Using the square root as a variance-stabilizing transformation in each case, we obtained the results indicated in table 5.24.

To construct $s_W^2$ and $s_B^2$ from the data, we note that

$$T_k = n_k \sqrt{\bar{\bar{X}}_k} \quad \text{and} \quad \sum_{i=1}^{n_k} (\sqrt{x_{ik}} - \sqrt{\bar{\bar{X}}_k})^2 = (n_k - 1)s_{\sqrt{x_k}}^2.$$

Then using expressions (9) and (12) we obtain

$$s_W^2 = \frac{1}{918} \sum_{k=1}^{15} (n_k - 1)s_{\sqrt{x_k}}^2 = 0.2778$$

and

$$s_B^2 = \frac{1}{14}\left[ \sum_{k=1}^{15} n_k (\sqrt{\bar{\bar{X}}_k})^2 - \frac{1}{933} \left( \sum_{k=1}^{15} n_k \sqrt{\bar{\bar{X}}_k} \right)^2 \right] = 3.4405.$$

The results of table 5.24 then yield

$$F = s_B{}^2/s_W{}^2 = 12.39$$

where the numerator has 14 and the denominator 918 degrees of freedom. From a table (e.g. table 18 in Pearson and Hartley 1962) of the $F$-distribution, we find that the 0.001 critical value for these degrees of freedom is

$$F_{14,918}^{0.001} \lesssim F_{12,\infty}^{0.001} = 2.74.$$

Thus

$$P(F \geq 2.74) \leq 0.001.$$

Hence our value of $F$ is highly significant, and we can reject the hypothesis of equal means. Thus the expected number of articles per passage cannot be assumed the same in these works.

In general the above procedure, called an *analysis of variance*, assumes as the null hypothesis that (i) each of the variates is normally distributed, (ii) they have a common variance, and (iii) they have a common mean. When this null hypothesis can be rejected, we would like to think it is due to the falsity of (iii) and not of either of the other assumptions. Fortunately, the procedure is much more sensitive to a violation of (iii) than to the violation of either (i) or (ii). Readers interested in details about just how much the actual probability of a type I error differs from the nominal significance levels when (i) and (ii) are violated should consult chapter 10 of Scheffé (1959).

### 5.4.5 *The Scheffé test for multiple comparisons*

From the results of example 4 we observe that the hypothesis

$$\mu_{\sqrt{x_1}} = \mu_{\sqrt{x_2}} = \cdots = \mu_{\sqrt{x_{15}}}$$

can be discarded because an analysis of variance yielded a significant $F$-value. Thus there appears to be a non-random variation in the rate of article use in the various sources. However, we do not know which differences among the individual means are significant.

We do, however, observe a certain difference between conversation samples and narrative samples as well as between samples that could be roughly classified as expository and those that could be classified as belles-lettres. If, for example, we wished to test the belles-lettres-expository opposition, then the most satisfactory procedure would be to obtain new data from the belles-lettres and the expository works and apply, for example, a $t$-test. However, when it is difficult or impossible, as it is in some cases, to obtain new data, the following test, due to Scheffé, can sometimes be used to answer such questions.

Consider a set of $r$ independent normally distributed random variables

$X_1, X_2, \ldots, X_n$ with common variance $\sigma_{X_k}^{\,2} = \sigma^2$. If a sample $(x_{1k}, x_{2k}, \ldots, x_{n_k k})$ is drawn from each $X_k$, then to test the $r(r-1)/2$ hypotheses

$$H_{kj}: \quad \mu_{X_k} = \mu_{X_j} \quad (k \neq j)$$

simultaneously, we use the following procedure. Form

$$(14) \quad F_{kj} = \frac{(\bar{X}_k - \bar{X}_j)^2}{s_W^{\,2}\left(\dfrac{1}{n_k} + \dfrac{1}{n_j}\right)(r-1)}$$

where $\bar{X}_k$ and $\bar{X}_j$ are the sample means of the $k$th and $j$th samples ($k, j = 0, 1, \ldots, r; k \neq j$), $s_W^{\,2}$ is given by expression (11), and

$$N = \sum_{k=1}^{r} n_k.$$

Then the probability that simultaneously all these $r(r-1)/2$ values of $F_{kj}$ are less than $F_{r-1,N-r}^{\alpha}$ is $1-\alpha$. Thus the probability of the complementary event, i.e., that $F_{kj} \geq F_{r-1,N-r}^{\alpha}$ for one or more of the $F_{kj}$, is $\alpha$. For a particular pair, say $\bar{X}_k$ and $\bar{X}_j$, the probability that

$$F_{kj} \geq F_{r-1,N-r}^{\alpha}$$

is less than or equal to $\alpha$. Therefore the probability of a type I error (rejecting the hypothesis when it is true) for any of the $r(r-1)/2$ comparisons does not exceed the significance level specified in an analysis of variance under the hypothesis

$$\mu_{X_1} = \mu_{X_2} = \ldots = \mu_{X_r}.$$

The probability of a type I error for some of these comparisons, however, may be considerably *lower* than the specified significance level. Thus the test is rather conservative.

Here we make a tacit assumption that if $F_{kj} \geq F_{r-1,N-r}^{\alpha}$, then the difference $\mu_{X_k} - \mu_{X_j}$ is non-zero and is one of the differences that contributes to the violation of the null hypothesis indicated in the original analysis of variance.

EXAMPLE 5   In example 4 we showed that among the 15 sources there was a significant variation in the mean number of articles per 50-word passage. Using

$$s_W^{\,2} = 0.2778$$

as obtained in example 4, we can form

$$F_{kj} = \frac{(\bar{X}_k - \bar{X}_j)^2}{(0.2778)\left(\dfrac{1}{n_k} + \dfrac{1}{n_j}\right)(14)} = (0.2585)\frac{n_k n_j (\bar{X}_k - \bar{X}_j)^2}{n_k + n_j}.$$

If we compare $W.C.$(conversation) with $W.C.$(narrative),[4] we find that

$$F = (0.2585) \frac{(32)\,(64)}{32+64} (1.57-2.21)^2 = 2.22$$

with 14 and 918 degrees of freedom, which is significant at the 0.025 level.

If we compare $L.J.$(conversation) with $L.J.$(narrative), we find a similar situation. In this case

$$F = (0.2585) \frac{(35)\,(35)}{35+35} (1.58-2.38)^2 = 2.90,$$

which is significant at the 0.001 level. Thus there is a significant difference in the use of articles in 'conversation' passages and narrative passages. The reader is invited to check the significance of some of the other

$$\frac{(15)\,(14)}{2} - 2 = 103$$

pairs.

The Scheffé test has a very useful refinement. Under the same hypothesis as is given above, if $c_1, c_2, \ldots, c_r$ are constants such that

$$\sum_{i=1}^{r} c_i = 0,$$

then the probability that

$$(15) \quad F = \frac{\left( \sum\limits_{i=1}^{r} c_i \bar{X}_i \right)^2}{s_W{}^2 \left( \sum\limits_{i=1}^{r} \frac{c_i{}^2}{n_i} \right) (r-1)} \geq F^{\alpha}_{r-1,\,N-r}$$

(where $s_W{}^2$ is given by expression (11)) is less than or equal to $\alpha$, where of course $F^{\alpha}_{r-1,\,N-r}$ is the $\alpha$-critical value of the $F$-distribution with $r-1$ degrees of freedom associated with the numerator and $N-r$ with the denominator.

The following example shows how (15) can be employed.

EXAMPLE 6   The items in table 5.24 can be divided into two classes: items 1–8 that correspond roughly to the category belles-lettres and items 9–15 that correspond roughly to expository (narrative) writing. Making the somewhat arbitrary choice that *The Autobiography of Bertrand Russell* is belles-lettres, we can use (15) to test whether the mean of the belles-lettres sample ($W.C.$ conversation, $S.M.$, $A.B.R.$, *Riders*, *Vic.*, *Nightmares*, $L.J.$, and $L.J.$ conversation) differs significantly

---

4   For a key to the abbreviations see the note to table 5.24 on page 201.

from the mean of the expository sample (*H.K.*, *Power*, *Observer*, *G.M.*, *W.C.* narrative, *T.P.H.*, and *L.J.* narrative). The belles-lettres mean is

$$\bar{X}_{BL} = \sum_{i \in BL} \frac{n_i \sqrt{\bar{X}_i}}{N_{BL}} = \sum_{i \in BL} \frac{n_i}{N_{BL}} \sqrt{\bar{X}_i} = 1.82$$

where $BL$ is the set of belles-lettres indices and

$$N_{BL} = \sum_{i \in BL} n_i = 467,$$

and the exposition mean is

$$\bar{X}_{EX} = \sum_{j \in EX} \frac{n_i}{N_{EX}} \sqrt{\bar{X}_i} = 2.18$$

where $EX$ is the set of expository indices and

$$N_{EX} = \sum_{\in EX} n_i = 466.$$

Then

$$\bar{X}_{EX} - \bar{X}_{BL} = \sum_{i=1}^{15} c_i \sqrt{\bar{X}_i}$$

where

$$c_i = n_i / N_{EX}$$

for $i \in EX$ and

$$c_i = -n_i / N_{BL}$$

for $i \in BL$. Then, using the results given in table 5.24, we can write (15) as follows:

$$F = \frac{(\bar{X}_{EX} - \bar{X}_{BL})^2}{s_W{}^2 \left(\dfrac{1}{N_{BL}} + \dfrac{1}{N_{EX}}\right)(14)} = \frac{(2.18 - 1.82)^2}{(0.2778)\left(\dfrac{1}{467} + \dfrac{1}{466}\right)(14)}$$

$$= 7.95.$$

Since $F_{14,918}^{0.001} = 2.74$, the probability that our $F$ is not less than 2.74 is itself less than or equal to 0.001. Our $F = 7.95$, so we can reject the hypothesis that the narrative mean and the belles-lettres mean are equal, with a type I error not exceeding 0.001.

EXERCISES

1 Find the 0.99 and 0.95 confidence limits of $\delta$ in example 2.

2 Using the data given in table 5.24, find the 0.95, 0.99, and 0.995 confidence limits of $\mu_{(n)}$, the mean of the distribution of the square root of the number of articles in 50-word passages of narrative in *The Wapshot Chronicle*, and of $\mu_{(c)}$, the mean of the distribution of passages containing conversation in the same work.

3 Using the data of table 5.24 and the *t*-test, test the difference between $\mu_{\sqrt{x_k}}$ and $\mu_{\sqrt{x_j}}$ for the values of $k$ and $j$ indicated: (a) 3, 5; (b) 4, 7; (c) 5, 9; (d) 5, 10; (e) 11, 9; (f) 9, 14.

4 Use the Scheffé test to test the significance of the pairs indicated in exercise 2.

5 Use the *F*-test to test the significance of the hypothesis that $\sigma_{\sqrt{x_{14}}}{}^2 = \sigma_{\sqrt{x_8}}{}^2$ in table 5.24. Why is this result not particularly damaging to the conclusion that we have drawn from the analysis of variance performed in example 5?

6 Among the sources in table 5.24 select the sources written by Bertrand Russell and apply an analysis of variance to test the hypothesis that they have a common mean value.

7 Apply the Scheffé test to the mean of the belles-lettres sample and the mean of the exposition sample as in example 6, but this time place *The Autobiography of Bertrand Russell* among the expository items and delete the conversation and narrative items.

## 5.5
### SOME MORE DISTRIBUTION-FREE TESTS

In sections 5.2 and 5.3 we considered a number of tests based on the $\chi^2$-distribution as well as the Kolmogorov test, none of which required any hypothesis about the underlying shape of the distributions of the random variables involved. The tests discussed in section 5.4, on the other hand, required that the random variables from which the samples were drawn be normally distributed, or at least approximately so. Now we return to consider more tests which make no requirement on the shape of the distribution of the random variable under consideration. Such tests are called *distribution-free* tests.

In this section we consider two of the many distribution-free tests. The first of these is for randomness.

### 5.5.1 *A runs-test for randomness*

Consider the set $S$ of sequences of the form $(x_1, x_2, ..., x_n)$ consisting of $n_0$ elements of one kind, say zeros, and $n_1$ of another, say ones. Thus each $x_i$ is either 0 or 1 and $n_0 + n_1 = n$. There are $\binom{n}{n_0}$ such sequences in $S$, and each of these sequences is composed of alternating *runs* of 0 and 1. The *length* of a run is the number of elements in it. If for $A \subseteq S$ we let

$$P(A) = \text{(number of elements in } A)\Big/ \binom{n}{n_0},$$

then $2^S$ and $P$ constitute a probability space in the sense of chapter 2, and if we let

$$R_i(x_1, x_2, \ldots, x_n) = \text{number of runs of } i$$

for $i = 0, 1$, then $R_0$ and $R_1$ are random variables defined on that probability space. It can be shown (Wilks 1962, p. 145, or Bradley 1968, p. 255) that

$$(1) \quad P(R_0 = r_0) = \binom{n_0-1}{n_0-r_0}\binom{n_1+1}{r_0}\Big/\binom{n}{n_0}.$$

An analogous result holds for $R_1$. A careful look at expression (1) shows that $R_0$ has the hypergeometric distribution (see section 3.1) with

$$P(R_0 = r_0) = h(r_0; n_0, n_1+1, n).$$

The formulas for the mean and the variance of the hypergeometric distribution are given in exercise 7 of section 3.2, and in the present case they take the form

$$(2) \quad \mu_{R_0} = \frac{n_0(n_1+1)}{n}$$

and

$$(3) \quad \sigma_{R_0}^2 = \frac{n_0 n_1(n_1+1)(n_0-1)}{n^2(n-1)}.$$

If $n_0$ and $n_1$ are large, then $R_0$ is approximately normally distributed with mean and variance given by (2) and (3) respectively. (See pp. 35–6 in Lancaster 1969; also Bradley 1968.)

The above development can be used as the basis for a test for randomness. Suppose each observation in a sequence $(x_1, x_2, \ldots, x_n)$ of $n$ observations results in one of two kinds of outcome $O_1$ and $O_2$. Suppose, in addition, that each sequence contains $n_0$ entries of the first kind and $n_1$ of the second. Under the null hypothesis that the sequence $(x_1, \ldots, x_n)$ of observations is a random sample from the population of all such sequences, the distribution of the number of runs is given by (1).

EXAMPLE 1 In sampling 1000 consecutive words from *The Solid Mandala* by Patrick White, it was observed that there were $n_0 = 710$ one-syllable words and $r_0 = 209$ runs of one-syllable words. Can this sample be treated as a random sample from a population with $n_0 = 710$ one-syllable and $n_1 = 1000-710 = 290$ polysyllable words? Under the null hypothesis of randomness, the number of runs of the first kind is given by (1), and, since $n_0$ and $n_1$ are large, we can assume that $R_0$ is normally distributed with mean

$$\mu_{R_0} = 206.61$$

and variance

$$\sigma_{R_0}^2 = 42.85.$$

Thus the deviation of our observed value of $R_0$ from the expected value $\mu_{R_0}$ is

$$R_0 - \mu_{R_0} = 209 - 206.61 = 2.39.$$

The probability of a greater deviation is given by the expression

$$P(|R_0 - \mu_{R_0}| \geq 2.39) = 2\{1 - N(\mu_{R_0} + 2.39, \mu_{R_0}, \sigma_{R_0})\}$$

$$= 2\left\{1 - N\left(\left|\frac{R_0 - \mu_{R_0}}{\sigma_{R_0}}\right|, 0, 1\right)\right\} = 2\left\{1 - N\left(\frac{2.39}{\sigma_{R_0}}, 0, 1\right)\right\}$$

$$= 2\{1 - N(0.3665, 0, 1)\} = 0.8557$$

which is certainly not significant. There appear to be neither significantly fewer nor significantly more runs than would be likely under the null hypothesis, so we have no evidence against the randomness hypothesis.

The distribution given by (1) can also be used in a wider context to test randomness. Suppose we have a sequence $(x_1, x_2, ..., x_n)$ of observations which we wish to test for randomness. We can convert the sequence into a sequence $(f(x_1), f(x_2), ..., f(x_n))$ of zeros and ones by choosing a value $c$, usually the median of the observations, and writing $f(x_i) = 0$ in the $i$th position of the associated sequence if $x_i \leq c$ and $f(x_i) = 1$ otherwise. The converted sequence can then be tested for randomness using (1).

EXAMPLE 2    In a given sample of article counts in fifty 50-word passages selected 'at random' (i.e., using a random-number table), is the hypothesis of actual randomness tenable? Fifty such passages from *A History of Western Philosophy* by Bertrand Russell were selected, and their article counts are listed below in the order of their selection:

    5, 4, 3, 5, 3, 4, 7, 5, 4, 4, 6, 4, 3, 9, 6, 5, 8, 5, 3, 4, 1, 6, 4, 4, 7, 5, 2, 7, 7, 4, 6,
    3, 4, 3, 5, 4, 4, 1, 7, 10, 2, 9, 8, 5, 7, 3, 2, 0, 5, 6.

The median of the sequence can be seen to be 4 in the sense that there are 25 observations less than or equal to 4. Thus if we let

$$f(x_i) = \begin{cases} 0 & \text{if } x_i \leq 4, \\ 1 & \text{otherwise,} \end{cases}$$

then the converted sequence of observations can easily be constructed. Using (1), the probability of $r_0$ runs of 0 is given by $P(R_0 = r_0)$ with $n_0 = n_1 = 25$. The observed number of runs of zero is $R_0 = 12$, which differs from the expected number

$$\mu_{R_0} = \frac{25 \times 24}{50} = 12$$

by zero. Hence there is no reason to doubt that the method of passage selection results in a genuinely random sample.

You may ask, however, whether it is possible that a lack of randomness might show up for another choice of $c$, say $c = 6$. If we let

$$f(x_i) = \begin{cases} 0 & \text{for } x_i \leq 6, \\ 1 & \text{otherwise,} \end{cases}$$

we find that the resulting sequence contains $n_0 = 40$ zeros and $n_1 = 10$ ones. The expected number of runs of zeros by (1) is

$$\mu_{R_0} = \frac{(40)(9)}{50} = 7.2,$$

and the observed number is $R_0 = 9$. The probability

$$P(|R_0 - \mu_{R_0}| \geq 1.8) = \sum_{r_0 \geq 9} P(R = r_0) + \sum_{R_0 \leq 5} P(R = R_0).$$

Since by (2)

$$\sigma_{R_0}{}^2 = \frac{(40)\,(10)\,(11)\,(39)}{(50)^2\,(49)} = 1.401$$

or $\sigma_{R_0} = 1.18$, the normal approximation yields

$$P(|R_0 - \mu_{R_0}| \geq 1.8) = 1 - [N(9 - \tfrac{1}{2}, 7.2, 1.18) - N(5 + \tfrac{1}{2}, 7.2, 1.18)]$$

$$= 1 - [N(1.10, 0, 1) - N(-1.44, 0, 1)]$$

$$= 1 - [N(1.10, 0, 1) - \{1 - N(1.44, 0, 1)\}]$$

$$= N(1.44, 0, 1) - N(1.10, 0, 1) = 0.9251 - 0.8643 = 0.0608,$$

which is not significant at the 0.05 level, so we still lack the evidence to reject the putative randomness. The reader should observe that since the sample values in the tails, or low-probability portions of the distribution, tend to occur sporadically, the power of this runs-test diminishes when $c$ is chosen too distant from the median. Also it should be observed that a choice of $c$ different from the median tends to skew the distribution of $R_0$ so that larger samples are required before the normal approximation becomes satisfactory.

### 5.5.2 *The Wilcoxon and related tests sensitive to unequal location*

Consider a random sample $(x_1, x_2, ..., x_n)$ of a random variable $X$ and a random sample $(y_1, y_2, ..., y_m)$ of a second random variable $Y$ independent of $X$. If these two random variables have identical distributions, then the combined sample $(x_1, x_2, ..., x_n, y_1, y_2, ..., y_m)$ is a sample from a random variable with the common distribution. Let the entries in this combined sample be ordered according to their magnitude in ascending (or descending) order to form an *ordered sample*

(4)  $(Z_{(1)}(w), Z_{(2)}(w), ..., Z_{(n+m)}(w))$

where

$$Z_{(i)}(w) \leqq Z_{(i+1)}(w)$$

for all $i = 1, 2, ..., n+m-1$ and $w$ is either $X$ or $Y$ according to whether $Z_i(w)$ is an observation of $X$ or $Y$. Assuming that there are no ties among the $Z_{(i)}(w)$, if $X$ and $Y$ are independent and identically distributed as we have assumed, then any assignment of $n$ $X$'s and $m$ $Y$'s to the $w$'s in

$$Z_{(1)}(w) < Z_{(2)}(w) < ... < Z_{(n+m)}(w)$$

is as likely to occur as any other. Since a particular assignment results from choosing a subset containing $n$ elements (to be assigned $X$) from the larger set containing $n+m$ elements, there are

$$\binom{n+m}{n} = \binom{n+m}{m}$$

possible assignments. Since these are equally likely, any one assignment occurs with probability

(5)  $P(Z_{(1)}(w), ..., Z_{(n+m)}(w)) = 1 \Big/ \binom{n+m}{n}.$

The tests we are about to discuss are based on the probability space associated with $2^S$ and $P$ where $S$ is the set of $\binom{n+m}{n}$ possible assignments of $X$'s and $Y$'s and $P$ is the probability function obtained from (5). These tests depend heavily on the assumption that there are no ties among the $Z_{(i)}(w)$. A small number of ties due to imprecision of measurement can be tolerated; however, the presence of ties decreases the power of the tests. If $X$ and $Y$ have a common *continuous* distribution, then, as was the case in the uniform distribution on [0, 1] where the probability that a particular point is chosen in that interval is zero, the probability of a tie is zero. It is possible that this condition of non-existence of ties could be met when the common distribution is discrete, but necessarily both $X$ and $Y$ must be capable of assuming a very large number of values compared with the sample size. As we shall soon see, relaxing this condition can lead to grave inaccuracies in the putative significance levels.

Since we are looking for a test sensitive to differences in location, we need to consider a statistic sensitive to such differences. Wilcoxon proposed the rank-sums

$$W(X) = \sum_{i=1}^{n} R_i(X), \qquad W(Y) = \sum_{i=1}^{m} R_i(Y)$$

where $R_i(X)$ is that $j$ in expression (4) such that $x_i = Z_{(j)}(X)$ and $R_i(Y)$ is that $j$ such that $y_i = Z_{(j)}(Y)$. For example, if the sample of $X$-values is (1.1, 2.1, 0.1, 0.05) and the sample of $Y$-values is (1.3, 0.13, 1.5, 1.6, 2.3), then $Z_{(1)}(X) = 0.05$, $Z_{(2)}(X) = 0.1$, $Z_{(3)}(Y) = 0.13$, $Z_{(4)}(X) = 1.1$, $Z_{(5)}(Y) = 1.3$, $Z_{(6)}(Y) = 1.5$, $Z_{(7)}(Y) = 1.6$, $Z_{(8)}(X) = 2.1$, and $Z_{(9)}(Y) = 2.3$. Thus $W(X)$, the sum of the $X$-ranks, is

$$W(X) = 1+2+4+8 = 15$$

and the sum of the $Y$-ranks is

$$W(Y) = 3+5+6+7+9 = 30.$$

It is clear in general that

$$W(X)+W(Y) = \sum_{i=1}^{n+m} i = \frac{(n+m)\,(n+m+1)}{2}$$

and that if $W(X)$, for example, is small with the ranks taken in ascending order, the $X$ values tend to be smaller than the $Y$. Tables of the lower tails of the $W$-distribution can be found in Bradley (1968, p. 318). These tables give the largest value of $w$ for which $P(W_n \leq w) \leq \alpha$ where $n$ is the size of the smaller sample.

For our example, the $X$-sample is smaller, so $n = 4$. Since the size of the other sample is $m = 5$, we look in the table headed $n = 4$ at the line corresponding to $m = 5$. Here we find that $0.005 \leq P(W_4 \leq 10) \leq 0.01$, $0.01 \leq P(W_4 \leq 11) \leq 0.025$, $0.025 \leq P(W_4 \leq 12) \leq 0.05$, and $0.05 \leq P(W_4 \leq 14) \leq 0.10$. Thus the probability that $W(X) \leq 15$ is greater than 0.10, and hence we could have obtained $W(X) = 15$ or less in more than 1 case in 10 under the hypothesis that $X$ and $Y$ have the same distribution.

The table considers values of $n$ from 1 to 25. For larger values of $n$

$$Z = \frac{W_n - [n(n+m+1)/2]}{\sqrt{nm(n+m+1)/12}}$$

has approximately the normal distribution $N(Z; 0, 1)$. To correct for continuity, the absolute value of the numerator must be reduced by $\frac{1}{2}$. Thus

$$P(W_n \leqq w) \approx N\left(\frac{w - \frac{1}{2} - [n(n+m+1)/2]}{\sqrt{nm(n+m+1)/12}}, 0, 1\right).$$

If the alternative to the null hypothesis is that the values of the sample with fewer observations tend to be larger than those of the other sample so that $W_n$ is near the upper tail, we can use the symmetry of the $W_n$-distribution about its mean $\overline{W}$ to obtain the probability that $W_n$ is greater than or equal to some number $w$ by observing that

$$P(w \leqq W_n) = P(W_n \leqq \overline{W} - (w - \overline{W})) = P(W_n \leqq 2\overline{W} - w).$$

Thus if we wish to obtain the $w$ for which $P(W_n \geqq w) \leqq \alpha$, we simply find the value in the table corresponding to $\alpha$, set it equal to $2\overline{W} - w$, and solve for $w$. The tables give $2\overline{W}$ for each pair of sample sizes. The following example illustrates this procedure.

EXAMPLE 3    Consider the set of means of $\sqrt{X_i}$, the square root of the article counts, given in table 5.24. These means might be construed as random variables themselves and we might test the hypothesis that the means for the belles-lettres items in the table have the same distribution as those for the expository items. Wilcoxon's test may be used to test this hypothesis. It is clear on linguistic grounds that the belles-lettres means should in general be smaller so we test the hypothesis of equal distribution against the hypothesis that the (distribution) mean of the belles-lettres distribution is smaller than the mean of the expository distribution.[5] Table 5.25 shows the ranks of the combined sample.

A glance at a table of the Wilcoxon distribution for $n = 5$ and $m = 10$ yields $2\overline{W} = 80$, so $W_5 \geq 56$ has the same probability as $W_5 \leqq 2\overline{W} - 56 = 80 - 56 = 24$. The table shows that

$$P(W_n \leqq 26) = P(W_n \geqq 54) \leqq 0.05$$

and 23 is the largest value such that

$$P(W_n \leqq 23) = P(W_n \geqq 57) \leqq 0.025.$$

Thus

$$0.025 \leqq P(W_n \geqq 56) \leqq 0.05,$$

and the difference between the belles-lettres observations and the expository observations in table 5.24 is significant at the 0.05 level using the Wilcoxon test. Compare this result with the results of example 6 of section 5.4.

The main use of the statistic $W_n$, and the Mann-Whitney statistic $U$ which we

5    However, were this not the case a priori, then a two-tailed test would be indicated.

TABLE 5.25

| Belles-lettres ranks | | Expository ranks | |
| --- | --- | --- | --- |
| 1 | *W.C.* (conversation) | | |
| 2 | *L.J.* (conversation) | | |
| 3 | *S.M.* | | |
| 4 | *A.B.R.* | | |
| 5 | *Riders* | | |
| 6 | *Vic.* | | |
| 7 | *Nightmares* | | |
| 8 | *L.J.* | | |
| | | 9 | *H.K.* |
| | | 10 | *Power* |
| | | 11 | *Obs.* |
| | | 12 | *G.M.* |
| 13 | *W.C.* (narrative) | | |
| | | 14 | *T.P.H.* |
| 15 | *L.J.* (narrative) | | |
| 64 | $m = 10$ | 56 | $n = 5$ |

shall discuss shortly, is in testing the hypothesis that two sets of test scores are from the same distribution. In this case it is assumed that there are sufficient questions and the graders are sufficiently incisive to obviate any possibility of tie scores. As mentioned above, a small number of ties is to be tolerated, in which case their collective ranks are averaged, but it should be borne in mind that ties do reduce the power of the test. The following example shows the problems that can arise when an excessive number of ties occur.

EXAMPLE 4    S.W. Becker and J. Carroll (1963) in one of their experiments tested 28 subjects, 13 of whom were asked to read a 10-sentence paragraph of low sentence contingency index (for a definition of this concept see the introduction to their paper) while the remaining 15 were given a comparable paragraph of high sentence contingency index. Shortly afterwards all 28 were given a seven-question multiple-choice test on the common subject matter of the paragraphs. The results of this test are given in table 5.26. The authors have averaged the ranks of the ties, which is a standard method of dealing with ties (Bradley 1968, pp. 50–4). It is easy to see from table 5.26 that there are an excessive number of ties, and we shall see that this distorts actual significance levels relative to the putative ones. If $X$ is the variate with the low-index scores and $Y$ the one with high-index scores, then, under the hypothesis of a common distribution for $X$ and $Y$,

$$W(X) = 137.5, \qquad W(Y) = 268.5,$$

and

$$0.005 \leqq P(W(X) \leqq 137.5) \leqq 0.01.$$

H

TABLE 5.26

| No. correct on the test | High index ranks (Y) | Low index ranks (X) |
|---|---|---|
| 1 | 4 | 2.5 |
| 2 | 4 | 2.5 |
| 3 | 4 | 2.5 |
| 4 | 4 | 2.5 |
| 5 | 5 | 7.5 |
| 6 | 5 | 7.5 |
| 7 | 5 | 7.5 |
| 8 | 5 | 7.5 |
| 9 | 5 | 7.5 |
| 10 | 5 | 7.5 |
| 11 | 6 | 17.5 |
| 12 | 6 | 17.5 |
| 13 | 6 | 17.5 |
| 14 | 6 | 17.5 |
| 15 | 6 | 17.5 |
| 16 | 6 | 17.5 |
| 17 | 6 | 17.5 |
| 18 | 6 | 17.5 |
| 19 | 6 | 17.5 |
| 20 | 6 | 17.5 |
| 21 | 6 | 17.5 |
| 22 | 6 | 17.5 |
| 23 | 6 | 17.5 |
| 24 | 6 | 17.5 |
| 25 | 7 | 26.5 |
| 26 | 7 | 26.5 |
| 27 | 7 | 26.5 |
| 28 | 7 | 26.5 |

Becker and Carroll used this result to say that the value $W(X) = 137.5$ was significant at approximately the 0.005 level. However, in this case we can actually obtain the exact probability $P(W(X) \leq 137.5)$. First note that the 13 $X$-values and 15 $Y$-values can be assigned to the 28 ranks of table 5.25 in $\binom{28}{13}$ ways. In the ranking obtained from Becker and Carroll's sample the first seven positions $(1, 2, ..., 7)$ contain $X$-values and the last 12 positions $(17, ..., 28)$ contain $Y$-values. Values of $W(X)$ less than or equal to 137.5, aside from the one given in table 5.26, can be obtained only by rearranging the three $X$-values and the six $Y$-values in the nine positions $(8, 9, ..., 16)$. The number of such arrangements, including that in table 5.26 is $\binom{9}{6}$. Therefore the number of assignments of 13 $X$-values and 15 $Y$-values which yield a $W(X) \leq 137.5$ is $\binom{9}{6}$, and the actual value of $P(W(X) \leq 137.5)$ is given by

$$P(W(X) \leq 137.5) = \binom{9}{6} \bigg/ \binom{28}{13} = \left(\frac{9!}{3! \, 6!}\right) \div \left(\frac{28!}{13! \, 15!}\right)$$

$$= \frac{3 \times 4 \times 7 \, (13!)(15!)}{28!} = \frac{3 \, (13!) \, (15!)}{(27)!}$$

$$= 3(6.22702 \times 10^9) \, (1.30767 \times 10^{12}) \, (0.918369 \times 10^{-28})$$

$$= 3(6.22702) \, (1.30767) \, (0.918369) \times 10^{-7} = 2.24 \times 10^{-6}.$$

Thus the putative value of the probability $P(W(X) \leq 137.5)$ given by the Wilcoxon test is of the order of $10^3$ times its actual value in this case. Therefore, in using the Wilcoxon test on this example, we are in grave danger of accepting a null hypothesis that is really false. Fortunately for Becker and Carroll, their result was putatively significant.

It has often been said that when the conditions for the application of both the Wilcoxon test and the $t$-test are satisfied, then they yield comparable results. Let us find $t$ in this case, where the conditions for neither test are fulfilled. For the two samples in table 5.26 we find that

$$\bar{X} = 5.15, \qquad \bar{Y} = 6.07,$$

$$s_X{}^2 = 0.808, \qquad s_Y{}^2 = 0.495,$$

$$n = 13, \qquad m = 15,$$

and so

$$t = 3.04$$

with $n + m - 2 = 26$ degrees of freedom. Since

$$P(t \geq 3.04) = 0.0025$$

for 26 degrees of freedom, the $t$-test gives an only slightly more reliable estimate of the true probability in this case.

In certain situations, like example 4, it is possible to obtain an exact value for the probability that the variate in question deviates by more than the sample value from the expected value. In such cases the true significance of the result can be obtained without the necessity of making the assumptions required in order to apply a particular test such as the Wilcoxon test in example 4.

The underlying probability space upon which the Mann-Whitney $U$-test is based is identical with the space $2^S$, $P$ upon which the Wilcoxon test is based. Indeed, $U_n$ is just a linear combination of $W_n$:

$$U_n = nm + \frac{n(n+1)}{2} - W_n,$$

and the two tests are therefore equivalent in their scope and results.

There are of course many useful and very subtle distribution-free, or non-parametric, tests in the statistical literature. However, it is surprising how often they are misused in practice. In order to obtain trustworthy results, it is always best to use a test only when the conditions of its application are known to be satisfied. For a readable and reliable account of a large number of these distribution-free tests, which contains explicit conditions for their application and interpretation as well as tables of their confidence levels, see Bradley (1968).

## EXERCISES

1 In 1000 consecutive words drawn from *Victory* by Joseph Conrad, it was observed that 682 were one-syllable words. In counting runs there were 220 runs of one syllable. Can this sample be treated as a random sample or is the order of occurrence of the number of syllables significant?

2 In a sequence of observations of 50-word blocks of text from *Riders in the Chariot* by Patrick White, the following pronoun counts were obtained: 4, 5, 4, 6, 2, 5, 4, 5, 3, 1, 4, 5, 6, 2, 2, 1, 2, 4, 6, 3, 1, 7, 9, 10, 4, 3, 4, 6, 1, 11, 9, 8, 8, 6, 4. Use the median test to test this sample for randomness.

3 Choose a book and count the number of articles in 20 successive blocks of 50 words each. Test your sample for randomness.

4 (a) Using example 3 as a model, test the hypothesis that the distribution of the belles-lettres means $\sqrt{\bar{X}_i}$ and that of the expository means in table 5.24 are identical when *A.B.R.* is assigned expository status. (b) Test the hypothesis that the two distributions are identical when only samples from complete works are considered, i.e., when the belles-lettres sample contains only items 3, 4, 5, 6, and 7 and the expository items contain only 8, 9, 10, 11, 12, and 14.

# 6
# Some more extended studies

In this chapter we shall treat some specific problems in more detail than we have done up to the present. This will allow us to synthesize the material developed in previous chapters as well as introduce one or two new topics based upon it.

We shall concentrate on three main areas: variations among pronouns and articles in literary texts, syllable counts in literary texts, and the use of the binomial distribution in lexicostatistical studies.

I should like to begin by making some general remarks on methodology. First, it goes without saying that what we observe and the manner in which we make our observations should be such as to yield trustworthy and indisuputable conclusions.

The use of statistics usually involves reducing the data to one or more numerical indices of description. If these indices are to fulfil the methodological goals of the previous paragraph they must at least satisfy the following conditions: (a) they must be objectively identifiable, so that different investigators will arrive at the same conclusion about the value of a particular index in a given set of data, and (b) they must be convenient to identify, so that the actual process of obtaining the index is not too complicated.

These conditions, if satisfied, should ensure that researchers can correctly assess the results obtained and build upon them, even at some time in the far distant future. In addition, if the present trend in the computerization of linguistic and literary research is to continue, as it surely will, it will be desirable to make the indices simple enough so as to base them on properties that are machine computable. To sum up, the measurement of these indices from the data should be *objective, routine, and mechanizable.*

Thus, in literary studies it is best to exhaust the usefulness of the easily measured indices such as article counts, pronoun counts, word-length counts, and sentence-length counts before proceeding to more complicated indices such as even noun counts (in English at least) where the identification either involves a great deal of care or is highly subjective or both. Thus, for example, the number of linguistic transformations applied to the deep structure of a sentence in order to arrive at its textual form (see the paper by Hayes in Doležel and Bailey 1969) is not a particularly productive index, for in present-day terms it is far from objective and certainly not routinely obtainable, and so it is hardly mechanizable.

Sometimes when a structure is not objectively definable, there exists a series of objectively definable indicators which when taken collectively are almost co-extensive with the given structure. These almost co-extensive indicators can be counted instead of the given structure in order to give an estimate of the number of times the given structure occurs. For example, inflected forms of the verb *to be* followed by a past participle could be taken as an indicator that a passive transformation had been applied. Although counting such a substitute would not yield an entirely accurate count of the number of times the passive transformation had been applied, it would yield a fairly good indication of it, with the advantages that it would be easier to count 'by hand' than counting the number of passive transformations applied and that it would be possible to count by computing machine in machine-readable text.

For the reasons given above, we have chosen in the next four sections to consider indices which are very straightforward in order to see how much information they will yield.

## 6.1

### A STUDY OF ARTICLE USE

The relative frequencies of individual words and of word classes have been used by various researchers as indicators of the style of certain authors, for example, Mosteller and Wallace (1964), and the style of certain periods (Hamilton 1949). The following study was undertaken in order to ascertain whether or not article frequency could be used as a style indicator. For another article study see J. Krámský (1967).

A naïve researcher might consider sampling the whole test of a number of works by various authors using various styles to obtain the relative frequency of articles in each case and then comparing the results. This process would be both costly in labour and wasteful of information for it would provide no measure of the variability of article use within the texts.

Another method, more to be recommended, would be to consider blocks of texts composed of, say, $n$ consecutive words. Suppose $m_i$ of these blocks are selected at random from a given text $T_i$ out of the collection of $q$ texts $T_1$, $T_2$, ..., $T_q$ under consideration. If $X_i$ stands for the number of articles in such an $n$-word block from text $T_i$, then each sample $(x_{1i}, x_{2i}, ..., x_{m_i i})$ of $X_i$ yields a sample histogram of the distribution of $X_i$ from which we can estimate the mean $\mu_i$ and variance $\sigma_i^2$ of $X_i$. The choice of $n$ is up to the researcher. It must be small enough so that those whose task it is to count the articles do not become fatigued in making the individual counts and commit errors, but it must also be large enough so that reasonable meaning can be attached to the results. In our study we chose $n = 50$. To test this choice, we ran a preliminary count of a hundred 50-word blocks from *Lord Jim* by Conrad to obtain a sample mean of

$\bar{X}_{50} = 4.44$ articles per 50 words and a sample variance of $s_{50}{}^2 = 4.794$, and a second preliminary count from the same work of forty 100-word blocks which resulted in a sample mean of $\bar{X}_{100} = 9.58$ articles per 100 words and a sample variance of $s_{100}{}^2 = 10.199$.

In the former case $\bar{X}_{50}/50$ is an estimate of the relative frequency $Y$ of articles in the work. Assuming that $\mu = E(Y) = E(X_{50}/50)$ is actually the relative frequency of articles in *Lord Jim*, we can use the results of theorem 1 in section 3.5 to estimate a confidence interval for $\mu$. From that theorem we have

$$(1) \quad P\left(\left|\frac{4.44}{50} - \mu\right| \geqq \varepsilon\right) \leqq \frac{4.794}{(50)^2}\frac{1}{100\varepsilon^2}$$

where $\sigma^2(X_{50}/50) = [1/(50)^2]\sigma^2(X_{50})$ by (13) in section 3.4 and $s_{50}{}^2$ is used as an estimate of $\sigma^2(X_{50})$. Choosing $\varepsilon$ so that the left side of (1) is 0.05, we find that $\varepsilon = 0.0196$ and

$$0.069 = 0.089 - 0.020 \leqq \mu \leqq 0.089 + 0.020 = 0.109$$

with probability 0.95. Using the second sample, if $\mu = E(Y) = E(X_{100}/100)$, then from theorem 1 of section 3.5

$$(2) \quad P\left(\left|\frac{9.58}{100} - \mu\right| \leqq \varepsilon\right) \leqq \frac{10.199}{(100)^2}\frac{1}{40\varepsilon^2}$$

where $s_{100}{}^2$ is used to estimate $\sigma^2(X_{100})$. If

$$\varepsilon = 0.0225,$$

then the probability that

$$0.073 = 0.096 - 0.023 \leqq \mu \leqq 0.096 + 0.023 = 0.118$$

is 0.95. Since these two intervals have approximately the same length and have a healthy overlap there appears to be little or no advantage to choosing the larger sample size if estimating $\mu$ is our only goal.

Table 6.1 contains the results of a survey of 22 sources. The number of random samples of 50-word blocks is indicated under each source in the row marked 'total'. In addition to the 15 sources included in table 5.24, the following seven sources were also included: (i) a second random sample from *Riders in the Chariot*, (ii) a second random sample from all of *Lord Jim*, (iii) a second random sample from *The Autobiography of Bertrand Russell* (A.B.R.), (iv) a random sample from the narrative passages of *A.B.R.*, (v) a random sample of the correspondence passages of *A.B.R.*, (vi) a random sample from *A History of Western Philosophy* (H.W.P.) by Bertrand Russell, and (vii) a random sample taken from the entire text of *The Wapshot Chronicle*.

The first question that might be asked about these data is: can a sample $(x_1, x_2, ..., x_{50})$ of 50 consecutive words be considered random in regard to the occurrence of articles? More specifically, when we replace $x_i$ by $u_i = 1$ if $x_i$ is an

TABLE 6.1*

| No. $i$ of articles in passage | S.M. | Riders (1st sample) | W.C. | W.C. (narrative) | W.C. (conversation) | L.J. (1st sample) | Vic. | Obs. | G.M. | H.K. | T.P.H. |
|---|---|---|---|---|---|---|---|---|---|---|---|
| No. $f_i$ of passages containing $i$ articles | | | | | | | | | | | |
| 0 | 6 | | 8 | 1 | 2 | 4 | 2 | | 1 | | |
| 1 | 13 | 3 | 8 | 2 | 4 | 5 | 4 | 2 | 1 | 3 | |
| 2 | 20 | 9 | 12 | 4 | 10 | 10 | 5 | 5 | 4 | 10 | |
| 3 | 24 | 12 | 11 | 7 | 7 | 12 | 9 | 13 | 16 | 13 | 7 |
| 4 | 17 | 8 | 14 | 12 | 4 | 22 | 10 | 10 | 25 | 20 | 14 |
| 5 | 11 | 8 | 4 | 10 | 3 | 18 | 8 | 16 | 18 | 13 | 12 |
| 6 | 6 | 8 | 8 | 9 | 1 | 13 | 5 | 14 | 15 | 9 | 18 |
| 7 | 2 | 1 | 8 | 10 | | 7 | 6 | 6 | 10 | 4 | 8 |
| 8 | 1 | 1 | 6 | 5 | 1 | 4 | | 3 | 6 | 4 | 7 |
| 9 | | | 2 | 2 | | 4 | | 1 | 4 | 3 | 1 |
| 10 | | | 2 | 2 | | 1 | | | | 1 | |
| 11 | | | | | | | 1 | | | | |
| TOTAL | 100 | 50 | 83 | 64 | 32 | 100 | 50 | 70 | 100 | 80 | 67 |
| MEAN $\bar{x}$ | 3.06 | 3.84 | 3.99 | 5.17 | 2.81 | 4.44 | 4.08 | 4.71 | 4.91 | 4.48 | 5.46 |
| VARIANCE $s^2$ | 3.007 | 2.831 | 7.134 | 4.875 | 2.996 | 4.794 | 4.647 | 3.106 | 3.416 | 4.075 | 2.374 |

* See table 5.24 (p. 201) for explanation of abbreviations.

TABLE 6.1 (continued)

No. $f_i$ of passages containing $i$ articles

| $i$ | Power | H.W.P. | A.B.R. (1st sample) | A.B.R. (2nd sample) | A.B.R. (narrative only) | A.B.R. (correspondence only) | Nightmares | Riders (2nd sample) | L.J. (2nd sample) | L.J. (narrative) | L.J. (conversation) |
|---|---|---|---|---|---|---|---|---|---|---|---|
| 0 |  | 1 | 2 | 3 |  | 1 | 2 |  | 1 |  | 1 |
| 1 |  | 2 | 5 | 3 | 3 | 7 | 1 | 3 | 3 | 1 | 8 |
| 2 | 5 | 3 | 8 | 14 | 6 | 10 | 8 | 7 | 7 | 2 | 7 |
| 3 | 8 | 7 | 12 | 12 | 14 | 12 | 7 | 7 | 5 | 3 | 9 |
| 4 | 9 | 12 | 11 | 2 | 9 | 9 | 11 | 11 | 13 | 2 | 3 |
| 5 | 12 | 9 | 6 | 9 | 8 | 7 | 7 | 9 | 7 | 6 | 7 |
| 6 | 11 | 5 | 5 | 5 | 6 | 3 | 7 | 9 | 6 | 4 |  |
| 7 | 3 | 6 |  | 2 | 3 |  | 2 | 1 | 2 | 10 |  |
| 8 |  | 2 | 1 |  |  |  | 5 | 2 | 1 | 5 |  |
| 9 | 2 | 2 |  |  | 1 | 1 |  | 1 | 3 | 1 |  |
| 10 |  | 1 |  |  |  |  |  |  | 1 |  |  |
| 11 |  |  |  |  |  |  |  |  |  | 1 |  |
| 12 |  |  |  |  |  |  |  |  | 1 |  |  |
| TOTAL | 50 | 50 | 50 | 50 | 50 | 50 | 50 | 50 | 50 | 35 | 35 |
| MEAN $\bar{x}$ | 4.70 | 4.76 | 3.38 | 3.28 | 3.98 | 3.22 | 4.26 | 4.26 | 4.52 | 5.89 | 2.74 |
| VARIANCE $s$ | 2.786 | 4.594 | 2.934 | 3.349 | 3.040 | 2.951 | 4.278 | 3.460 | 6.175 | 4.751 | 2.256 |

article and $u_i = 0$ if $x_i$ is not an article, is the resulting sequence $(u_1, u_2, ..., u_{50})$ a random sample from a population containing

$$u = \sum_{i=1}^{50} u_i$$

ones and $50 - u$ zeros? The runs-test developed in section 5.5 might be used to try to answer this question. However, a moment's consideration indicates that, since articles cannot occur in consecutive positions, the number of runs of articles will equal exactly the number of articles among the $x_i$. Thus randomness cannot be assumed in order to obtain a model of this situation that in turn can be used to obtain a theoretical distribution of the variate $X_i$, the number of articles per 50-word passage in text $T_i$.

However, observation of the data indicates that the Poisson distribution (see section 3.3) might fit the data in some cases and the translated Poisson distribution (see section 5.2) might fit the data in others. Table 6.2 indicates the result of applying a $\chi^2$-distribution to ascertain the goodness-of-fit of these distributions to the data in various cases.

Except for *The Wapshot Chronicle* and *The Theory and Processes of History*, the agreement is adequate. Indeed, the agreement is very good in all other cases except for *Power* and the newspapers, where the agreement is only fair.

Notice that *The Theory and Processes of History* fits a Poisson distribution translated three units to the right and appears to be anomalous in the uniformly high article count that it exhibits. (None of the 67 blocks counted contains fewer than three articles.) *Power* is the only one of the remaining sources that comes near this record.

*The Wapshot Chronicle* is anomalous in that it is a mixture of different types

TABLE 6.2

| Source | Sample size | Mean | Theoretical distribution | $\chi^2_{obs}$ | DF | To nearest 10th $P(\chi^2 > \chi^2_{obs})$ |
|---|---|---|---|---|---|---|
| *S.M.* | 100 | 3.06 | Poisson | 1.01 | 6 | 0.9856 |
| *Riders* (1st sample) | 50 | 3.84 | Poisson | 1.79 | 4 | 0.7725 |
| *L.J.* (1st sample) | 100 | 4.44 | Poisson | 3.92 | 7 | 0.7798 |
| *Victory* | 50 | 4.08 | Poisson | 1.53 | 5 | 0.9013 |
| *H.K.* | 80 | 4.48 | Poisson | 3.75 | 6 | 0.7306 |
| *T.P.H.* | 67 | 5.46 | Poisson | 15.25 | 5 | ≈0.009 |
| *T.P.H.* | 67 | 5.46 | 3-trans. Poisson | 3.15 | 4 | 0.5578 |
| *W.C.* (total) | 83 | 3.99 | Poisson | 27.00 | 6 | 0.00015 |
| *W.C.* (narrative) | 60 | 5.17 | Poisson | 2.17 | 6 | 0.9004 |
| *G.M.* | 100 | 4.91 | Poisson | 8.15 | 7 | 0.3326 |
| *Obs.* | 70 | 4.71 | Poisson | 7.08 | 5 | 0.2206 |
| *Power* | 50 | 4.70 | Poisson | 6.82 | 5 | 0.2360 |
| *Power* | 50 | 4.70 | 2-trans. Poisson | 5.65 | 5 | 0.3471 |
| *A.B.R.* (narrative) | 50 | 3.98 | Poisson | 3.21 | 5 | 0.6692 |

of writing. Some passages are diary entries, the author signalling this style by a drastic reduction in article use; others are strongly narrative in quality, yielding a high average ($\bar{X} = 5.17$) use of article; and still others are conversational with a moderately low article usage ($\bar{X} = 2.81$). Oddly enough, *Lord Jim* exhibits a similar contrast between conversation and narrative passages, but this does not seem to affect the Poisson quality of the over-all *Lord Jim* sample.

As an experiment, we tried to fit a mixed Poisson distribution to *The Wapshot Chronicle* data. Consider that the work is made up of three stores $S_1$, $S_2$, $S_3$ of 50-word passages corresponding to diary, conversation, and narrative passages respectively. Assuming that the number of articles in the passages in each store has the Poisson distribution with parameters $\lambda_1$, $\lambda_2$, $\lambda_3$ respectively, and that the relative frequency in the whole text of passages from $S_i$ is $\beta_i$, then

$$(3) \quad P(X = k) = \sum_{i=1}^{3} P(\text{passage sampled belongs to } S_i) \cdot P(X = k \mid \text{passage in } S_i)$$

$$= \sum_{i=1}^{3} \beta_i \frac{\lambda_i^k e^{-\lambda_i}}{k!}$$

for $k = 0, 1, 2, \ldots$ where

$$(4) \quad \beta_1 + \beta_2 + \beta_3 = 1.$$

There are essentially five parameters to be evaluated in order to obtain a completely determined expression for $P(X = k)$. The method of moments would require the calculation of five moments in order to obtain these parameters. Since we already estimated from other samples that $\bar{X}$ for conversational passages is 2.81 and for narrative passages 5.17, we take a sample of diary passages to obtain a mean of 0.90 and then hypothesize that $\lambda_1 = 0.90$ for diary passages, $\lambda_2 = 2.81$ for conversational passages, and $\lambda_3 = 5.17$ for narrative passages. Then expression (3) takes the form

$$P(X = k) = \beta_1 \frac{(0.90)^k e^{-0.90}}{k!} + \beta_2 \frac{(2.81)^k e^{-2.81}}{k!} + \beta_3 \frac{(5.17)^k e^{-5.17}}{k!}.$$

Thus we need only estimate $\beta_1$ and $\beta_2$. Remember that $\beta_3 = 1 - \beta_1 - \beta_2$ by definition. Using the method of moments, we obtain the equations

$$(5) \quad \mu_X = \sum_{k=0}^{\infty} k P(X = k) = \beta_1 \lambda_1 + \beta_2 \lambda_2 + \beta_3 \lambda_3$$

and

$$(6) \quad E(X(X-1)) = \sum_{k=0}^{\infty} k(k-1) P(X = k) = \beta_1 \lambda_1^2 + \beta_2 \lambda_2^2 + \beta_3 \lambda_3^2.$$

Since

$$E(X(X-1)) = E(X^2) - E(X)$$

and

$$\sigma_X{}^2 = E[(X - E(X))^2] = E(X^2) - [E(X)]^2,$$

we obtain from (4), (5), and (6) the set of equations

$$\beta_1 + \beta_2 + \beta_3 = 1,$$

$$\beta_1 \lambda_1 + \beta_2 \lambda_2 + \beta_3 \lambda_3 = \bar{X},$$

$$\beta_1 \lambda_1{}^2 + \beta_2 \lambda_2{}^2 + \beta_3 \lambda_3{}^2 = \frac{n-1}{n} s_X{}^2 - \bar{X} + \bar{X}^2$$

for the method-of-moments estimates of the $\beta_i$. Using the $\lambda_i$ hypothesized above and the $\bar{X}$ and

$$s_X{}^2 = \frac{1}{n-1} \sum_{i=1}^{n} (x_i - \bar{X})^2$$

for the general sample from *The Wapshot Chronicle* (shown in table 6.1) in these equations, we obtain the following estimates for the $\beta_i$:

$$\beta_1 = 0.2043, \qquad \beta_2 = 0.1312, \qquad \beta_3 = 0.6645.$$

With these estimates we can compute the expected number of articles predicted by our model. In table 6.3 the observed results given in table 6.1 are shown opposite the results predicted by the model.

A test for goodness-of-fit yields $\chi^2 = 7.33$ with seven degrees of freedom, nine from the table minus two because of the calculation of the estimates for $\beta_1$ and $\beta_2$ from the sample. $P(\chi^2 \geq 7.4) = 0.3885$, so our result does not depart significantly from that predicted by the model. The values of the $\beta_i$ could be checked by taking a random sample from the book to obtain other estimates of the relative frequencies of the three kinds of passage.

If we leave aside *The Wapshot Chronicle* and *The Theory and Processes of History*, the remaining nine sources considered in table 6.2 have a definite Poisson character. The other 11 sources considered in table 6.1 resemble sources considered in table 6.2 to a sufficient extent that we are assured of their Poisson character also.

Since our reason for undertaking a study of articles was to ascertain whether a variation in their usage could be considered an indication of differences in genre or authorship, we would like to find a statistic $\tau$ for which it is possible to

TABLE 6.3

| No. of articles | 0 | 1 | 2 | 3 | 4 | 5 | 6 | 7 | 8 | 9 | 10 |
|---|---|---|---|---|---|---|---|---|---|---|---|
| Observed | 8 | 8 | 12 | 11 | 14 | 4 | 8 | 8 | 6 | 2 | 2 |
| Predicted | 7.86 | 9.67 | 9.57 | 10.49 | 11.23 | 10.64 | 8.77 | 6.32 | 4.03 | 4.42 | |

judge whether differences in its values for various works are statistically significant or not, i.e., whether the values could or could not (the latter indicating significance) have occurred by chance. The commonest statistic used for this purpose is the sample mean. However, it is an unfortunate fact that there has been very little sampling theory developed for the mean of a Poisson variate. Nevertheless all is not lost. It is known that if $X$ is a Poisson variate with distribution mean in the range of values we are considering here, then $\sqrt{X}$ is approximately normally distributed with variance $\sigma_{\sqrt{X}}^2$ near $\frac{1}{4}$. This matter, already considered in section 5.4, is developed in Kendall and Stuart (1966, pp. 88–90). If $X_i$ is the number of articles in a 50-word passage, then the values of $n_i$, the sample size, $\sqrt{\overline{X_i}}$, the mean of the square roots of the sample values, and $s_{\sqrt{X_i}}$ are given in table 6.4 for the sources indicated there.

In order to test whether the two samples for *Riders in the Chariot*, *Lord Jim*, and *The Autobiography of Bertrand Russell* are significantly different, and hence sufficiently different to preclude their combination into one sample, we apply the *t*-test outlined in section 5.4.

A.B.R., 1st sample versus 2nd sample:

$$t = \frac{1.75 - 1.71}{\sqrt{\left(\frac{49(0.5635)^2 + 49(0.6171)^2}{98}\right)\left(\frac{1}{50} + \frac{1}{50}\right)}} = 0.3977$$

with 98 degrees of freedom. A glance at a table of the *t*-distribution shows that

$$0.6 \leq P(|t| \geq 0.397) \leq 0.7.$$

*Riders*, 1st sample versus 2nd sample:

$$t = 1.1055$$

with 98 degrees of freedom. In this case

$$0.2 \leq P(|t| \geq 1.106) \leq 0.3.$$

*Lord Jim*, 1st sample versus 2nd sample:

$$t = 0.2108$$

with 148 degrees of freedom;

$$0.8 \leq (P|t| \geq 0.211) \leq 0.9.$$

Since none of these results are at all significant, there is no reason not to combine the two samples in each case into a single sample. The resulting means and variances are given in lines 21, 22, and 23 of table 6.4. Since the newspaper samples resemble each other closely ($t = 0.56$ with 168 degrees of freedom so that $0.5 \leq P(|t| \geq 0.56) \leq 0.8$), we also combine them. The results of the combination are given in line 24 of table 6.4.

TABLE 6.4

| $i$ | Name | $n_i$ | $\sqrt{\bar{X}_i}$ | $s_{\sqrt{x_i}}$ |
|---|---|---|---|---|
| 1 | *W.C.* (conversation) | 32 | 1.57 | 0.5979 |
| 2 | *L.J.* (conversation) | 35 | 1.58 | 0.5182 |
| 3 | *S.M.* | 100 | 1.64 | 0.6044 |
| 4 | *A.B.R.* (2nd sample) | 50 | 1.71 | 0.6171 |
| 5 | *A.B.R.* (correspondence) | 50 | 1.72 | 0.5215 |
| 6 | *A.B.R.* (1st sample) | 50 | 1.75 | 0.5635 |
| 7 | *Riders* (1st sample) | 50 | 1.91 | 0.4380 |
| 8 | *Victory* | 50 | 1.92 | 0.6346 |
| 9 | *A.B.R.* (narrative) | 50 | 1.95 | 0.4477 |
| 10 | *Nightmares* | 50 | 1.97 | 0.6108 |
| 11 | *Riders* (2nd sample) | 50 | 2.01 | 0.4740 |
| 12 | *L.J.* (1st sample) | 100 | 2.01 | 0.6351 |
| 13 | *L.J.* (2nd sample) | 50 | 2.03 | 0.6280 |
| 14 | *H.K.* | 80 | 2.06 | 0.4823 |
| 15 | *H.W.P.* | 50 | 2.11 | 0.5657 |
| 16 | *Power* | 50 | 2.13 | 0.4016 |
| 17 | *Obs.* | 70 | 2.13 | 0.4302 |
| 18 | *G.M.* | 100 | 2.17 | 0.4608 |
| 19 | *W.C.* (narrative) | 64 | 2.21 | 0.5616 |
| 20 | *L.J.* (narrative) | 35 | 2.38 | 0.4981 |
| | | | | |
| 21 | *A.B.R.* (sum of two samples) | 100 | 1.73 | 0.5884 |
| 22 | *Riders* (sum of two samples) | 100 | 1.96 | 0.4602 |
| 23 | *L.J.* (sum of two samples) | 150 | 2.02 | 0.6307 |
| 24 | *Obs.+G.M.* | 170 | 2.15 | 0.4476 |

In order to test whether the over-all difference in means among the samples is significant, we could perform a one-way analysis of variance of items 1, 2, 3, 5, 21, 8, 9, 22, 10, 23, 14, 15, 16, 24, 19, and 20 of table 6.4. However, in section 5.4 we performed an analysis of variance on approximately this same material (see table 5.24) with highly significant results, so it seems reasonable to infer significant differences among the means in this case too. In the exercises the reader is asked to perform the analysis of variance on the items listed above to see if our inference is indeed correct. He will find that under the null hypothesis of equal means, the quotient of the average sum of squares between the samples over the average sum of squares within the samples is 11.05 with 15 degrees of freedom for the numerator and 1150 for the denominator, which is significant at the 0.005 level.

Some readers may feel that it is questionable whether we are justified in applying an analysis of variance on the items listed above and by extension the items listed in table 5.24 because the range

$$s_{\sqrt{x_{16}}}{}^2 = (0.4016)^2 = 0.1613 \leqq s_{\sqrt{x_i}}{}^2 \leqq 0.4027 = (0.6346)^2 = s_{\sqrt{x_8}}{}^2$$

of estimates of the putative common variance is too great. However, as we have stated in section 5.4, the analysis of variance is not sensitive to slight departures from the hypothesis of a common variance. See Scheffé (1959, section 10.4) for a discussion of the effects of departure from the underlying assumptions in this case and others.

To see how serious the departure from the assumption of equal variance is, we can apply *Bartlett's test for equal variance*, which is as follows. If $\{s_k^2 \mid k = 1, ..., m\}$ is a set of unbiased sample variances obtained from $m$ independent random samples containing $n_1, n_2, ..., n_m$ observations respectively from normal populations with common variance $\sigma^2$, then

$$(7) \quad M = N \ln \left( \frac{1}{N} \sum_{k=1}^{m} v_k s_k^2 \right) - \sum_{k=1}^{m} v_k \ln(s_k^2),$$

where

$$v_k = n_k - 1, \qquad N = \sum_{i=1}^{k} v_k,$$

and $\ln x$ stands for the natural logarithm of $x$, has approximately the $\chi^2$-distribution with $m-1$ degrees of freedom.

If we construct $M$ for the 16 items listed above, we obtain

$$M = -1414.38 + 1460.01 = 45.63$$

and with 15 degrees of freedom we find that

$$P(\chi^2 \geq 44) = 0.00011,$$

which indicates a significant departure from the equality of variances.

It is interesting to note that, if we leave aside item 24, then among the list of remaining items under consideration (1, 2, 3, 5, 8, 9, 10, 14, 15, 16, 19, 20, 21, 22, and 23) there are 10 items (1, 2, 3, 5, 8, 10, 15, 19, 21, and 23) whose sample contains at least one block with zero article frequency, and that this set of items is also characterized by the fact that they are exactly the items in the list with variance greater than or equal to $(0.5182)^2 = 0.2685$. If we apply Bartlett's test to these 10 items, we find that

$$M = -698.84 + 703.86 = 5.015$$

with nine degrees of freedom, and

$$P(\chi^2 \geq 5) = 0.83,$$

which is evidence for a common variance among these 10 items. The range of sample variances in this group is

$$(8) \quad 0.2685 \leq s_{\sqrt{x_i}}^2 \leq 0.4027.$$

If we take the remaining six items in our list (including item 24), we note that the range of their variances is

(9)  $0.1613 \leqq s_{\sqrt{X_i}}^{2} \leqq 0.2481$

and that for them

(10) $M = -753.45 + 756.15 = 2.70$

with five degrees of freedom. Thus with $P(\chi^2 \geqq 2.6) = 0.76$ we have evidence for a common variance among these items.

Assuming that there are two variances $\sigma_1^2$ and $\sigma_2^2$, the former common to the samples with sample variances in the range (8) and the latter with sample variances in the range (9), and assuming that the actual values of these variances lie centred within the ranges of their respective sample variances, then the ratio $\sigma_1^2/\sigma_2^2$ of these variances is less than 2, which is not great enough to have more than a mild effect on the results of an analysis of variance.

If we list the means of $\sqrt{X_i}$ for the items under consideration in increasing order of magnitude, we find

$$\sqrt{\overline{\overline{X_1}}} = 1.57, \quad \sqrt{\overline{\overline{X_2}}} = 1.58, \quad \sqrt{\overline{\overline{X_3}}} = 1.64, \quad \sqrt{\overline{\overline{X_5}}} = 1.72,$$

$$\sqrt{\overline{\overline{X_{21}}}} = 1.73, \quad \sqrt{\overline{\overline{X_8}}} = 1.92, \quad \sqrt{\overline{\overline{X_9}}} = 1.95, \quad \sqrt{\overline{\overline{X_{22}}}} = 1.96,$$

$$\sqrt{\overline{\overline{X_{10}}}} = 1.97, \quad \sqrt{\overline{\overline{X_{23}}}} = 2.02, \quad \sqrt{\overline{\overline{X_{14}}}} = 2.06, \quad \sqrt{\overline{\overline{X_{15}}}} = 2.11,$$

$$\sqrt{\overline{\overline{X_{16}}}} = 2.13, \quad \sqrt{\overline{\overline{X_{24}}}} = 2.15, \quad \sqrt{\overline{\overline{X_{19}}}} = 2.21, \quad \sqrt{\overline{\overline{X_{20}}}} = 2.38.$$

The only conspicuous gaps occur between $\sqrt{\overline{\overline{X_{21}}}}$ and $\sqrt{\overline{\overline{X_8}}}$, where the difference is 0.19, and between $\sqrt{\overline{\overline{X_{19}}}}$ and $\sqrt{\overline{\overline{X_{20}}}}$, where the difference is 0.17; in all other cases the successive differences are less than 0.07.

A somewhat dubious procedure would be to consider items 1, 2, 3, 5, and 21 together and run a one-way analysis of variance upon their article counts under the null hypothesis of equal means. If we do this, we obtain the following within-column and between-column variances:

$s_W^2 = 0.3333$   with 312 degrees of freedom,

$s_B^2 = 0.2913$   with four degrees of freedom.

These yield

$$F_{312}^{4} = s_B^2/s_W^2 = 0.8739$$

which, being only slightly less than one, indicates a non-significant deviation from the expected value, 1.

If we group the remaining items together and perform another analysis of variance, we obtain

$s_W^2 = 0.2770$   with 838 degrees of freedom,

$s_B^2 = 0.9832$   with ten degrees of freedom.

Thus

$$F^{10}_{838} = 3.549,$$

which is significant at the 0.001 level.

If we discard item 20 and include only the 10 items 8, 9, 22, 10, 23, 14, 15, 16, 24, and 19, we obtain

$$s_W{}^2 = 0.2783 \quad \text{with 804 degrees of freedom,}$$

$$s_B{}^2 = 0.7122 \quad \text{with nine degrees of freedom.}$$

Thus

$$F^9{}_{804} = 2.559,$$

so

$$0.005 \leqq P(F^9_{804} \geqq 2.56) \leqq 0.01,$$

which is still significant at the 0.01 level. However, since we are categorizing the data a posteriori and treating them more than once, the levels of significance in the tables will be higher than the actual levels.

A more satisfactory procedure might be to use some method of multiple comparison such as Scheffé's test developed in section 5.4. The mean sum of squares within the samples in the total list of $r = 16$ items is

$$s_W{}^2 = 0.2923,$$

and there are $N = 1166$ counts in all; thus

$$F^{15}_{1150} = \frac{(\sqrt{\overline{\overline{X}}_i} - \sqrt{\overline{\overline{X}}_j})^2}{0.2923 \left( \dfrac{1}{n_i} + \dfrac{1}{n_j} \right) 15} .$$

If we take a critical level of 0.05, then $F^{15}_{1150} = 1.67$ and the difference between $\sqrt{\overline{\overline{X}}_i}$ and $\sqrt{\overline{\overline{X}}_j}$ is significant if

$$(11) \quad |\sqrt{\overline{\overline{X}}_i} - \sqrt{\overline{\overline{X}}_j}| \geqq 2.706 \sqrt{\dfrac{1}{n_i} + \dfrac{1}{n_j}} .$$

If both $n_i$ and $n_j$ are 50, then (11) becomes

$$|\sqrt{\overline{\overline{X}}_i} - \sqrt{\overline{\overline{X}}_j}| \geqq 0.541.$$

If one is 100 and the other 50, then (11) becomes

$$|\sqrt{\overline{\overline{X}}_i} - \sqrt{\overline{\overline{X}}_j}| \geqq 0.469.$$

The largest difference among items with Russell's authorship, for example, is $\sqrt{\overline{X}_{16}} - \sqrt{\overline{X}_5} = 0.408$, which is subcritical. However, Scheffé's test is known to have diminished power in the sense that the probability of accepting the null hypothesis when it is false is abnormally high. Because of this we make an analysis of variance of the Russell items and compare the result with the above. Among the seven Russell items,

$$s_W{}^2 = 0.2782 \quad \text{with 423 degrees of freedom,}$$

$$s_B{}^2 = 1.8812 \quad \text{with six degrees of freedom,}$$

and so

$$F_{423}^6 = 6.7621,$$

which is highly significant, the 0.001 level being approximately 3.74. Because we are treating the same data a number of times, there is a multiple comparison effect, and the probability of rejecting a hypothesis when it is true is increased slightly, but the value of $F$ in this case is so large that it is reasonable to infer significance at the 0.05 level at least.

Since little is known about the power of Scheffé's test, we might tend to place more confidence in the analysis-of-variance result and infer a significant difference among the means of the Russell works.

If we make a comparison between *Victory* and *Lord Jim*, the two Conrad works, we find

$$t = 0.948 \quad \text{with 198 degrees of freedom,}$$

which is non-significant. However, a comparison between *Riders in the Chariot* and *The Solid Mandala* yields

$$t = 4.17 \quad \text{with 198 degrees of freedom,}$$

and so nominally

$$P(|t| \geqq 4.0) = 0.00006,$$

which should be small enough to offset the multiple comparison effect. The reader may also object that the sample variances are sufficiently different to indicate a difference in population variances. However, the $t$-test is insensitive to differences in variance when the sample sizes are equal (see Scheffé, 1959, p. 340).

One lesson to be drawn from these considerations is that Scheffé's test is useful when it yields nominally significant results but is unreliable when the results are not nominally significant. On the other hand, when multiple comparisons are involved, repeated use of analysis of variance can diminish the actual significance levels relative to the nominal levels.

With these considerations in mind, we can make the following conclusions about the data given in table 6.4.

1. The article count in 50-word blocks of text appears from this data not to be author-specific. Indeed we have found that the square roots of the article

counts in *The Solid Mandala* and *Riders in the Chariot* differ significantly and the averages of the square roots of article counts in the works of Russell surveyed also indicate significant differences, and finally, although the over-all counts for *Victory* and *Lord Jim* showed no significant difference, we find that the differences between conversation and narrative in *Lord Jim* is

$$|\sqrt{X_2} - \sqrt{X_{20}}| = 0.801.$$

The Scheffé test shows that this difference is critical because differences greater than

$$2.706 \sqrt{\frac{1}{32} + \frac{1}{35}} = 0.662$$

are significant. Thus article use in different types of passage in the same work of Conrad also shows critical differences. The same is true of the two types of passage in *The Wapshot Chronicle*.

2. The article count does appear from this study to be sensitive to differences in kinds of writing. Newspaper writing, narrative, and formal expository writing tend to employ more articles in general than novels, correspondence, and, if this sample is any indication, autobiography. (See also section 5.4, example 6, and section 5.5, example 3.) However, a great deal more work must be done to bring these rather vague outlines of results into sharper focus.

## EXERCISES

1 Perform the $\chi^2$-test of goodness-of-fit on the 10 sources of table 6.1 that are not represented in table 6.2, using the theoretical class estimates given by the Poisson distribution with $\lambda = \bar{X}$ in each case.

2 Solve equations (5). Most desk computers have package programs for solving three linear equations with three unknowns. One of these can be used, or the reader can calculate the result 'by hand,' referring to an introductory book on linear algebra should this be required. See, for example, Brainerd et al. (1967, volume 4, section 5.2).

3 Perform a one-way analysis of variance on items 1, 2, 3, 5, 21, 8, 9, 22, 10, 23, 14, 15, 16, 24, 19, and 20 in table 6.4. Use as a model the analysis of variance performed on table 5.24 in chapter 5.

## 6.2
A CONSIDERATION OF PRONOUNS VERSUS ARTICLES

In some cases among the data collected in section 6.1, the number of tokens of the personal pronouns *I, me, my, mine, you, your, yours, ..., we, us, our ours, ..., they, them, their, theirs* in 50-word blocks of text was obtained simultaneously

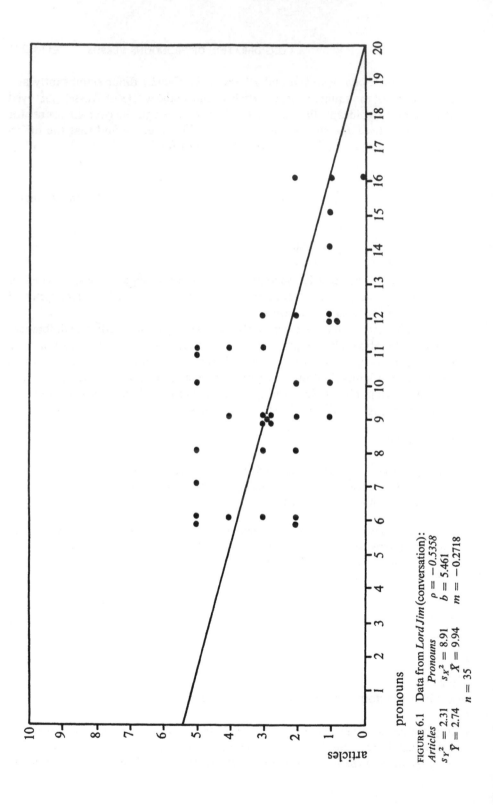

pronouns

FIGURE 6.1  Data from *Lord Jim* (conversation):

| *Articles* | *Pronouns* | $\rho = -0.5358$ |
| --- | --- | --- |
| $s_Y^2 = 2.31$ | $s_X^2 = 8.91$ | $b = 5.461$ |
| $\bar{Y} = 2.74$ | $\bar{X} = 9.94$ | $m = -0.2718$ |
| | $n = 35$ | |

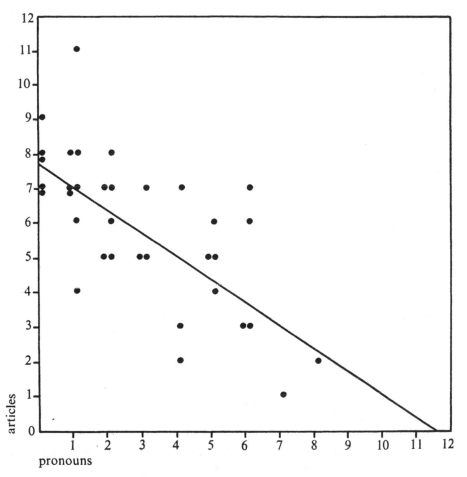

FIGURE 6.2   Data from *Lord Jim* (narrative):

| Articles | Pronouns | $\rho = -0.6987$ |
|---|---|---|
| $s_Y^2 = 4.75$ | $s_X^2 = 5.24$ | $b = 7.787$ |
| $\bar{Y} = 5.89$ | $\bar{X} = 2.86$ | $m = -0.6651$ |
| | $n = 35$ | |

with the article-token count, and in others a new random sample was taken to obtain these quantities. Thus the sample taken in each case was composed of pairs $(x_i, y_i)$ where $x_i$ indicates the number of pronouns in the $i$th block of the sample and $y_i$ the number of articles in that block. In order to see whether it is possible to posit a relationship between the number of pronouns and the number of articles, we plotted some of the samples on graph paper as in figures 6.1–6.4. The reader will note that at least a qualitative connection between the number of articles and the number of pronouns in a passage appears to be present

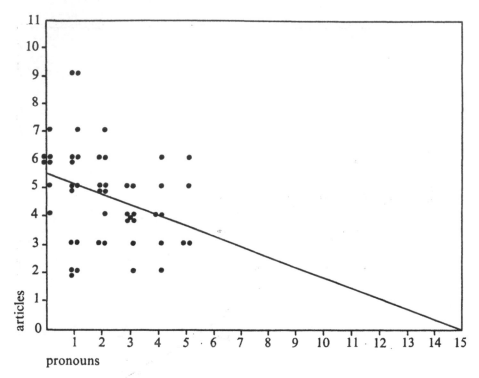

FIGURE 6.3   Data from *Power*:

| Articles | Pronouns | $\rho = -0.3216$ |
|---|---|---|
| $s_Y^2 = 2.79$ | $s_X^2 = 2.26$ | $b = 5.450$ |
| $\bar{Y} = 4.70$ | $\bar{X} = 2.10$ | $m = -0.3575$ |
| | $n = 50$ | |

in these figures. In particular, when the number of articles in a block is large the number of pronouns is small and vice versa.

In section 1.4 we developed the method of linear regression which fitted the best line, $y = mx+b$, to a scatter of sample points, in the sense that if $\{(x_i, y_i)|\ i = 1, 2, ..., n\}$ is the set of sample points, then $m$ and $b$ are chosen so that the 'distance'

$$d = \sqrt{\sum_{i=1}^{n} [y_i-(mx_i+b)]^2}$$

from the sample to the line is as small as possible. As a by-product of this procedure, we can compute

$$\rho_{XY} = \frac{1}{n-1} \sum_{i=1}^{n} \frac{(x_i-\bar{x})\,(y_i-\bar{y})}{s_X s_Y},$$

the sample correlation coefficient which we found in section 1.4 to be a measure

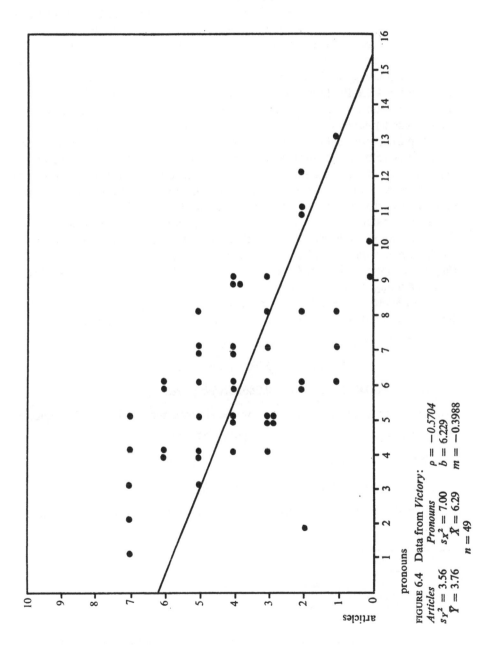

pronouns

FIGURE 6.4  Data from *Victory*:

| *Articles* | *Pronouns* | $\rho = -0.5704$ |
|---|---|---|
| $s_Y{}^2 = 3.56$ | $s_X{}^2 = 7.00$ | $b = 6.229$ |
| $\bar{Y} = 3.76$ | $\bar{X} = 6.29$ | $m = -0.3988$ |
| | $n = 49$ | |

articles

of how well the line approximated the set of sample points. If $m > 0$, then larger values of $\rho_{XY}$ (i.e., near 1) indicates a good approximation, and if $m < 0$, then values of $\rho_{XY}$ near $-1$ indicate a good approximation. Values of $\rho_{XY}$ near 0 indicate a poor approximation and a tendency toward independence between $X$, the number of pronouns used, and $Y$, the number of articles used.

For $X =$ the number of pronouns in a 50-word block and $Y =$ the number of articles in the same block, the values of $m$, $b$, $\rho_{XY}$, $s_X^2$, $\bar{X}$, $s_Y^2$, $\bar{Y}$, and $n$ (the sample size) are given in table 6.5 for the 15 samples indicated. The lines in figures 6.1–6.4 are the lines $y = mx+b$ where, in each case, $m$ and $b$ are chosen from the appropriate row of table 6.5. These lines in each of the 15 cases indicate the average tendency of pronouns to 'replace' articles in the text.

A partial explanation of this tendency might lie in the facts that (i) pronouns (especially in the third person) replace noun phrases, (ii) some pronouns (e.g., *my*) replace articles, and (iii) articles are, in some general sense, indicators of noun phrases in the text. On the other hand, some pronouns (e.g., *I*, *me*, and perhaps *you*) do not, in general, replace noun phrases.

If we consider only random samples from entire works and look at works by the same author, we learn some interesting things. The values of $\rho_{XY}$, $m$, and $b$ for *The Solid Mandala* and *Riders in the Chariot* by Patrick White and for *Lord Jim* and *Victory* by Joseph Conrad are quite similar within the works of each author. Among the works of Bertrand Russell sampled, *Power* and *A History of Western Philosophy* resemble each other closely, but the parameters in the other three Russell items vary somewhat more. Over the five Russell samples,

$$(A.B.R.) \; -0.55 \; \leqq \rho \leqq \; -0.31 \; (H.K.),$$

$$(Power) \quad 5.4507 \leqq b \leqq \quad 5.9318 \; (H.W.P.),$$

$$(H.K.) \quad -0.4720 \leqq m \leqq \; -0.2862 \; (Nightmares),$$

compared with a variation of these parameters over the whole set of samples of

$$(L.J. \text{ narr.}) \; -0.70 \quad \leqq \rho \leqq \; -0.29 \quad (T.P.H.),$$

$$(S.M.) \qquad 4.4274 \leqq b \leqq \quad 7.7859 \; (L.J. \text{ narr.}),$$

$$(L.J. \text{ narr.}) \; -0.6651 \leqq m \leqq \; -0.2389 \; (S.M.).$$

Note that the differences between the highest and lowest values arrange themselves as follows:

|            | Russell | Total |
|------------|---------|-------|
| $\Delta\rho$ | 0.24    | 0.41  |
| $\Delta b$   | 0.48    | 3.36  |
| $\Delta m$   | 0.18    | 0.43  |

Thus, except for values of $b$, Russell samples are somewhat diffuse relative to the

TABLE 6.5

|  | $m$ | $b$ | $\rho_{XY}$ | $\bar{X}$ | $s_X^2$ | $\bar{Y}$ | $s_Y^2$ | $n$ |
|---|---|---|---|---|---|---|---|---|
| T.P.H. | −0.3034 | 5.8563 | −0.293 | 1.87 | 2.009 | 5.29 | 2.157 | 38 |
| Obs. | −0.3107 | 5.4645 | −0.390 | 2.41 | 4.884 | 4.71 | 3.106 | 70 |
| S.M. | −0.2389 | 4.4274 | −0.403 | 5.91 | 7.557 | 3.02 | 2.767 | 100 |
| Riders | −0.2262 | 4.9910 | −0.393 | 5.62 | 8.240 | 3.72 | 2.736 | 50 |
| Vic. | −0.3988 | 6.2291 | −0.570 | 6.29 | 7.000 | 3.76 | 3.564 | 49 |
| L.J. | −0.4084 | 7.0717 | −0.579 | 6.13 | 9.250 | 4.57 | 4.557 | 98 |
| L.J. (conversation) | −0.2718 | 5.4610 | −0.536 | 9.94 | 8.906 | 2.77 | 2.307 | 34 |
| L.J. (narrative) | −0.6651 | 7.7859 | −0.699 | 2.86 | 5.244 | 5.89 | 4.751 | 35 |
| A.B.R. | −0.3855 | 5.5155 | −0.554 | 5.54 | 6.049 | 3.38 | 2.934 | 50 |
| A.B.R. (correspondence) | −0.3741 | 5.4121 | −0.559 | 5.86 | 6.572 | 3.22 | 2.951 | 50 |
| A.B.R. (narrative) | −0.3326 | 5.7896 | −0.479 | 5.44 | 6.292 | 3.98 | 3.040 | 50 |
| Nightmares | −0.2862 | 5.4650 | −0.415 | 4.18 | 9.008 | 4.26 | 4.278 | 50 |
| H.W.P. | −0.3639 | 5.9318 | −0.354 | 3.22 | 4.338 | 4.76 | 4.594 | 50 |
| Power | −0.3575 | 5.4507 | −0.322 | 2.10 | 2.255 | 4.70 | 2.786 | 50 |
| H.K. | −0.4720 | 5.6840 | −0.312 | 2.60 | 1.646 | 4.63 | 2.145 | 40 |

total spread. However, if we were to discard the *Lord Jim* narrative sample, which appears to be an outlier, we should obtain the following extremes for parameters:

$$(L.J.) \quad -0.58 \; \leqq \rho \leqq \; -0.29 \quad (T.P.H.),$$

$$(S.M.) \quad 4.4274 \leqq b \leqq \quad 7.0717 \, (L.J.),$$

$$(H.K.) \; -0.4720 \leqq m \leqq \; -0.2389 \, (S.M.),$$

which yields a new set of total-sample differences

$$\Delta \rho = 0.29, \quad \Delta b = 2.67, \quad \Delta m = 0.23.$$

Thus the spread in the Russell samples for $\rho$ and $m$ appears to be almost representative of the 14 remaining samples when the *Lord Jim* narrative sample is removed.

The distinct lack of variation in $b$ for the Russell works indicates that the datum value of the number of articles used when no pronouns appear (which is in some sense what $b$ is) seems to vary comparatively little in the works considered. It might be of interest to look into this question in a wider sample of his works.

Now let us proceed to a somewhat more detailed study of these data using the Wilcoxon test developed in section 5.5.[1] The 11 samples in table 6.5 may be

---

1 Because of the possible lack of normality in the distribution of $(X, Y)$, we are avoiding the use of the well-developed hypothesis-testing theory available for the coefficients $\rho$, $m$, and $b$ in the standard texts. See, for example, Williams (1959), or Chakravarti et al. (1967, especially section 7.9), or for that matter any of the standard introductory texts on statistical analysis. Rather, we push our comparisons as far as we can using non-parametric methods.

divided into two parts. The first (group I), containing *S.M.*, *Vict.*, *L.J.*, *Nightmares*, *Riders*, and *A.B.R.*, can be characterized roughly as belles-lettres. The remaining five items (group II), *Power*, *H.W.P.*, *H.K.*, *Observer*, and *T.P.H.*, might be characterized as exposition.

For each of the parameters $\rho$, $b$, and $m$, the Wilcoxon test can be used to test whether the values in group I differ significantly from the values in group II. In this case the null hypothesis is that all the values of $m$, for example, obtained in both I and II are drawn from the same (null) distribution. Because the sample size among the various works is not uniform, it might be a contributing factor to any significance obtained in the test. Therefore what follows should be taken as preliminary. For a more thorough-going study using the Wilcoxon test it would be advisable to keep the sample size uniform throughout.

From table 6.6 it is clear that the rank-sums for the values of $m$ from I and II are $W_6(\text{I}) = 39$ and $W_5(\text{II}) = 27$, which are not at all significant. Thus $m$, which may be taken to be the average rate of replacement of articles by pronouns, is not sensitive to the difference between belles-lettres and exposition as we have interpreted the terms here.

The reader should note that even if we place *A.B.R.* in the exposition column, which some readers might feel is more appropriate, then $W_6(\text{I})$ is changed to $W_5(\text{I}) = 35$ and $W_5(\text{II})$ to $W_6(\text{II}) = 31$, which is even less of a deviation from what is to be expected under the null hypothesis.

On the other hand, if we leave *A.B.R.* among the belles-lettres and discard the Conrad entries on the grounds that Conrad, not being a native speaker of English, will have an anomalous style, the belles-lettres ranks become 9 (*Riders*), 8 (*S.M.*), 7 (*Nightmares*), and 2 (*A.B.R.*). Thus the sum of the group I ranks is $W_4(\text{I}) = 26$, and there are $m = 5$ entries in group II. If we test the null hypothesis of equal distributions against the hypothesis of a larger distribution mean for the

TABLE 6.6

| Work | Value of $m$ | Belles-lettres ranks (I) | Exposition ranks (II) |
|------|------|------|------|
| *Riders* | −0.2262 | 11 | |
| *S.M.* | −0.2389 | 10 | |
| *Nightmares* | −0.2862 | 9 | |
| *T.P.H.* | −0.3034 | | 8 |
| *Obs.* | −0.3107 | | 7 |
| *Power* | −0.3575 | | 6 |
| *H.W.P.* | −0.3639 | | 5 |
| *A.B.R.* | −0.3855 | 4 | |
| *Vic.* | −0.3988 | 3 | |
| *L.J.* | −0.4084 | 2 | |
| *H.K.* | −0.4720 | | 1 |
| | | 39 | 27 |

$m$-values of I than those in II, then $2\overline{W}_4 - W_4 = 14$ and so

$$0.05 \leqq P(2\overline{W}_4 - W_4 \leqq 14) = P(W_4 \geqq 26) \leqq 0.1$$

is obtained from the tables. Thus there is not quite significance at the 0.05 level.

If, however, $A.B.R.$ is also moved to the exposition column, in addition to the changes mentioned in the previous paragraph, then the sum of the group I ranks becomes 24, and $2\overline{W}_3 - W_3 = 6$, so we find that

$$P(2\overline{W}_3 - W_3 \leqq 6) = P(W_3 \geqq 24) \leqq 0.025,$$

which is significant. There is thus some evidence that the values of $m$ for belles-lettres items tend to be higher than those for expository writing (including biography), provided of course that the works of Conrad are excluded.

The ranks for the $b$-values are given in table 6.7. In this case $W_5(\text{II}) = 31$, which is clearly not significant ($\overline{W}_5 = 30$ when there are six entries in the other sample). However, Conrad's works again provide the anomaly. If we delete Conrad's works, we find $W_4(\text{I}) = 14$ with $m = 5$, so

$$0.05 \leqq P(W_4 \leqq 14) \leqq 0.10,$$

which is almost significant again at the 0.05 level. If we move $A.B.R.$ over to the exposition column, we find $W_3(\text{I}) = 8$ with $m = 7$. Thus from the table of the Wilcoxon distribution

$$0.025 \leqq P(W_3 \leqq 8) \leqq 0.05,$$

which again proves to be significant.

If the works are ranked with respect to $\rho_{XY}$, then it is clear from table 6.8 that the values of the belles-lettres correlations are in general smaller than those of the exposition correlations. We have carried $\rho_{XY}$ to three decimal places to

TABLE 6.7

| Work | Value of $b$ | Belles-lettres ranks (I) | Exposition ranks (II) |
|------|------|------|------|
| S.M. | 4.4274 | 1 | |
| Riders | 4.9910 | 2 | |
| Power | 5.4507 | | 3 |
| Obs. | 5.4645 | | 4 |
| Nightmares | 5.4650 | 5 | |
| A.B.R. | 5.5155 | 6 | |
| H.K. | 5.6837 | | 7 |
| T.P.H. | 5.8563 | | 8 |
| H.W.P. | 5.9318 | | 9 |
| Vic. | 6.2291 | 10 | |
| L.J. | 7.0717 | 11 | |
| | | 35 | 31 |

TABLE 6.8

| Work | Value of $\rho_{XY}$ | Belles-lettres ranks (I) | Exposition ranks (II) |
|---|---|---|---|
| *L.J.* | −0.579 | 1 | |
| *Vic.* | −0.570 | 2 | |
| *A.B.R.* | −0.554 | 3 | |
| *Nightmares* | −0.415 | 4 | |
| *S.M.* | −0.403 | 5 | |
| *Riders* | −0.393 | 6 | |
| *Obs.* | −0.390 | | 7 |
| *H.W.P.* | −0.354 | | 8 |
| *Power* | −0.322 | | 9 |
| *H.K.* | −0.312 | | 10 |
| *T.P.H.* | −0.293 | | 11 |

obviate the tie between *Riders* and *Observer*. Note here that the Conrad works do not rank differently from the other belles-lettres samples.

For more information on the processing of binary data, and regression analysis in particular, see Cox (1970) and Williams (1959).

## 6.3
### ON THE DISTRIBUTION OF SYLLABLES PER WORD IN ENGLISH TEXTS [2]

The distribution of syllables per word in a corpus is (in European languages at least) an easily accessible index of its style, and it has been suggested by at least one author (Elderton 1949) that it might be an index by which one can distinguish one author from another. Before we can test its value in distinguishing among authors, and what is more likely among genres, it might be useful to investigate some of its general properties.

Among the questions that can be asked are the following: (i) Does the distribution of syllables per word in a given work conform to any of the well-known probability distributions? (ii) Does the distribution remain constant throughout a single work or the works of a single author? (iii) Can a sample of consecutive words in a text be taken as a random sample of syllables per word within the text? (iv) Does the distribution of syllables per word for a specific part of speech, say nouns, in a text differ from that for the text as a whole?

We shall consider each of these questions in turn.

### 6.3.1 *The distribution of syllables per word*

The distribution of the number of syllables per word in a text has been considered by various authors. The first to consider this point was W.P. Elderton (1949),

2  This section is a slight revision of a note appearing under the same title in *Keiryo Kokugo Gaku* (*Mathematical Linguistics*), no. 57(1971), 1–18.

who hypothesized that it might follow a form of geometric distribution but found the agreement rather poor. Next W. Fucks (1955) hypothesized that the translated Poisson distribution was the proper distribution. Fucks, together with Moreau (1964), Brandt Corstius (1970), and Zierer (1964), tested this distribution against data in German, French, Dutch, and Spanish with nearly universally negative results. Since English was not among the languages considered in these studies, we feel bound to test this distribution on English data.

Consider a corpus $w_1 w_2 \ldots w_m$ composed of word-tokens $w_i$ and a random variable $X$ such that $X(w_i)$ is the number of syllables in $w_i$. Fucks's suggestion is that

$$(1) \quad P(X = k) = \frac{(E(X) - 1)^{k-1} e^{-[E(X) - 1]}}{(k-1)!}$$

for $k = 1, 2, 3, \ldots$ where $E(X)$ stands for the expected value or distribution mean of $X$.

As an initial assumption, we assume that a sample of $n$ consecutive words from the corpus under consideration constitutes a random sample. We shall consider the adequacy of this assumption in detail later.

We sampled $n$ consecutive words from nine sources: *Riders in the Chariot* (*Riders*) by Patrick White ($n = 500$ consecutive words), *The Wapshot Chronicle* (*W.C.*) by John Cheever ($n = 1007$ consecutive words), *A Portrait of the Artist as a Young Man* (*Portrait*) by James Joyce ($n = 539$ consecutive words), *The Nigger of the Narcissus* (*N.N.*) by Joseph Conrad ($n = 2000$ consecutive words), *Julius Caesar* (*J.C.*) ($n = 500$, with 100 consecutive words from each act) and *As You Like It* (*A.Y.L.I.*) ($n = 500$, with 100 consecutive words from each act), both by Shakespeare, *The Lady of the Lake* (*L.L.*) by Walter Scott ($n = 500$, with 100 consecutive words each from cantos 1, 2, 3, 4, and 6), *The Cocktail Party* (*C.P.*) by T.S. Eliot ($n = 500$ consecutive words from act III), and *Time* of 8 June 1970 ($n = 500$ consecutive words). From these samples we obtained the observed (obs.) frequencies of words of various syllables as given in table 6.9. Using the sample mean as an estimate of $E(X)$, we obtain the corresponding expected (exp.) frequencies under the null hypothesis that the distribution of $X$ is given by (1). The $\chi^2$-goodness-of-fit test yields the $\chi^2$-values marked at the base of each column. Note that not only must the column of expected frequencies add up to $n$ (thus accounting for the loss of one degree of freedom), but also we are using the sample mean in our construction of the expected value (which accounts for the loss of another degree of freedom). Thus there are two fewer degrees of freedom than there are cells in the column of each work.

The results of the table indicate that the translated Poisson distribution yields a poor fit in general for the works considered. With the exception of *The Lady of the Lake*, these results conform to the patterns obtained for other languages (see references listed above).

Since the data in Elderton (1949, p. 441) for *Henry IV*, part 2, appear to be

TABLE 6.9

| | Riders | | W.C. | | Portrait | | N.N. | | J.C. | | A.Y.L.I. | | L.L. | | C.P. | | Time | |
|---|---|---|---|---|---|---|---|---|---|---|---|---|---|---|---|---|---|---|
| $\bar{x}$ | 1.4060 | | 1.3873 | | 1.3859 | | 1.3865 | | 1.2900 | | 1.2720 | | 1.2580 | | 1.4260 | | 1.7340 | |
| $s$ | 0.6915 | | 0.6943 | | 0.6449 | | 0.7248 | | 0.5988 | | 0.5715 | | 0.4857 | | 0.7783 | | 0.9492 | |
| Syllables per word | Obs. | Exp. | Obs. | Exp. | Obs. | Exp. | Obs. | Exp. | Obs. | Exp. | Obs. | Exp. | Obs. | Exp. | Obs. | Exp. | Obs. | Exp. |
| 1 | 347 | 333.16 | 713 | 683.64 | 375 | 366.43 | 1441 | 1358.86 | 388 | 374.13 | 393 | 380.93 | 382 | 386.30 | 357 | 326.56 | 268 | 239.99 |
| 2 | 111 | 135.26 | 221 | 264.77 | 124 | 141.41 | 400 | 525.20 | 84 | 108.50 | 81 | 103.61 | 107 | 99.66 | 93 | 139.11 | 131 | 176.15 |
| 3 | 35 | 27.46 | 54 | 51.27 | 36 | 27.28 | 119 | 101.49 | 23 } | | 23 } | | 11 } | | 30 | 29.63 | 74 | 64.55 |
| 4 | 6 } | | 15 } | | 4 } | | 28 } | | 5 } | 17.37 | 3 } | 15.46 | 0 } | 14.04 | 20 } | | 20 } | |
| 5 | 1 } | 4.13 | 4 } | 7.31 | 0 } | 3.88 | 10 } | 14.44 | 0 } | | 0 } | | 0 } | | 0 } | 4.70 | 7 } | 19.20 |
| ≥6 | 0 } | | 0 } | | 0 } | | 2 } | | 0 } | | 0 } | | 0 } | | 0 } | | 0 } | |
| TOTAL | 500 | | 1007 | | 539 | | 2000 | | 500 | | 500 | | 500 | | 500 | | 500 | |
| DF | 2 | | 2 | | 2 | | 2 | | 1 | | 1 | | 1 | | 2 | | 2 | |
| $\chi^2$ | 8.99* | | 8.64* | | 5.13 | | 83.08**** | | 12.55**** | | 12.50*** | | 1.25 | | 67.93**** | | 19.36*** | |

* Significant at 0.05 level.    $\bar{x}$ = sample mean.
*** Significant at 0.001 level.    $s$ = sample standard deviation.
**** Significant at 0.0001 level.

TABLE 6.10
Fitting the translated Poisson distribution to
Elderton's data

|  | | *Henry* IV, pt. 2 (prose) | | *Henry* IV, pt. 2 (poetry) | |
|---|---|---|---|---|---|
|  | $\bar{x}$ | 1.2517 | | 1.2772 | |
|  | $s$ | 0.5674 | | 0.5854 | |
| Syllables per word | | Obs. | Exp. | Obs. | Exp. |
| 1 | | 10,965 | 10649.89 | 9,076 | 8778.03 |
| 2 | | 2,177 | 2680.58 | 1,918 | 2433.27 |
| 3 | | 430 | 337.35 | 476 | 337.25 |
| 4 | | 99 | 28.30 | 108 | 31.16 |
| 5 | | 23⎫ | 1.88 | 4⎫ | 2.29 |
| $\geqq 6$ | | 4⎭ | | 0⎭ | |
| TOTAL | | 13,698 | | 11,582 | |

complete, the results of a translated Poisson fit given in table 6.10 are particularly damaging to the Fucks theory. An inspection of the table indicates that a $\chi^2$-text would yield a $\chi^2$-value vastly greater than the degree of freedom.

As observed by Moreau (1964; see also example 3 in section 5.2), in the case of French, the deviation of the expected frequencies from the observed is quite consistent from item to item (with only one exception, *The Lady of the Lake*) in that its estimate for the frequencies of one-syllable words is too low, for two-syllable words too high, and for the remainder too low. Thus, if we are going to obtain a better fit, we must find a distribution that concentrates more probability on one-syllable words. One such possibility is the translated negative binomial distribution[3] with

$$(2) \quad P(X = m) = p^k \binom{k}{m-1} (1-p)^{m-1}$$

for $m = 1, 2, 3, \ldots$ where $k$ and $p$ are real parameters $(0 < k, 0 \leqq p \leqq 1)$.

Estimates of the values of the parameters $k$ and $p$ can be obtained from the method of moments where

$$(3) \quad E(X) = 1 + \frac{k(1-p)}{p}$$

and

$$(4) \quad \sigma_X{}^2 = \frac{E(X)-1}{p} .$$

---

3   It is of interest to note that if $k = 1$ this distribution reduces to the geometric distribution, Elderton's original conjecture concerning the distribution of syllables per word. Thus our conjecture of the translated negative binomial distribution is a generalization of Elderton's conjecture. For more information on other possible distributions, see Patil and Joshi (1968).

Thus

(5) $\quad \hat{p} = \dfrac{\bar{X}-1}{s_X{}^2} \quad$ and $\quad \hat{k} = \dfrac{(\bar{X}-1)\hat{p}}{1-\hat{p}}$

are the method-of-moments estimates of $p$ and $k$ respectively. Sometimes a better estimate of $p$ and $k$ can be achieved by using only the first two sample frequencies $f_1$ and $f_2$, where $f_i$ stands for the sample frequency of $i$-syllable words, as follows. Choose estimates $p$ and $k$ such that

(6) $\quad f_1 = N\hat{p}^{\hat{k}}$

and

(7) $\quad f_2 = N\hat{p}^{\hat{k}}\hat{k}(1-\hat{p})$.

Since this involves solving a pair of transcendental equations, we proceed to use (3) and the ratio of (7) and (6), i.e., (3) and

(8) $\quad f_2/f_1 = \hat{k}(1-\hat{p})$.

These equations yield the estimates

(9) $\quad \hat{p} = \dfrac{f_2}{f_1(\bar{X}-1)} \quad$ and $\quad \hat{k} = \dfrac{f_2}{f_1(1-\hat{p})}$.

The various values of $\hat{p}$ and $\hat{k}$ for both methods of estimation are given in table 6.11. One can see immediately that *The Lady of the Lake* sample yields an impossible value for $p$ because $p \leq 1$ by the definition of the negative binomial distribution. Shortly we shall consider more samples from this work in order to see whether the sample in table 6.9 is representative of the poem as a whole. For *Riders*, *W.C.*, *Portrait*, and *N.N.* the estimates obtained by either method are quite similar, but for the others (except for *L.L.*) the estimates from the two

TABLE 6.11

| | Method of moments | | Using $f_1$ and $f_2$ | |
|---|---|---|---|---|
| | $\hat{p}$ | $\hat{k}$ | $\hat{p}$ | $\hat{k}$ |
| *Riders* | 0.8491 | 2.2839 | 0.7879 | 1.5082 |
| *W.C.* | 0.8076 | 1.6340 | 0.8003 | 1.5522 |
| *Portrait* | 0.9279 | 4.9646 | 0.8569 | 2.3107 |
| *N.N.* | 0.7357 | 1.0760 | 0.7182 | 0.9850 |
| *J.C.* | 0.8088 | 1.2267 | 0.7465 | 0.8540 |
| *A.Y.L.I.* | 0.8328 | 1.3548 | 0.7577 | 0.8506 |
| *L.L.* | 1.0937 | −3.0126 | 1.0857 | −3.2693 |
| *C.P.* | 0.7033 | 1.0096 | 0.6115 | 0.6705 |
| *Time* | 0.8147 | 3.2270 | 0.6659 | 1.4631 |
| *H.* IV, pt. 2 (prose) | 0.7818 | 0.9018 | 0.7888 | 0.9401 |
| *H.* IV, pt. 2 (poetry) | 0.8088 | 1.1728 | 0.7624 | 0.8893 |

TABLE 6.12

| | Method of moments estimate | | Estimate using $f_1$ and $f_2$ | |
|---|---|---|---|---|
| | $\chi^2$ | DF | $\chi^2$ | DF |
| *Riders* | 1.69 | 1 | 1.69 | 1 |
| *W.C.* | 1.20 | 2 | 0.31 | 2 |
| *Portrait* | 2.99* | 1 | 3.04* | 1 |
| *N.N.* | 0.87 | 3 | 0.81 | 3 |
| *J.C.* | 1.21 | 1 | 0.80 | 1 |
| *A.Y.L.I.* | 2.68 | 1 | 2.37 | 1 |
| *C.P.* | 4.97** | 1 | 0.81 | 1 |
| *Time* | 7.02** | 2 | 9.04** | 2 |
| *H.* IV, pt. 2 (prose) | 1.17 | 2 | 1.36 | 2 |
| *H.* IV, pt. 2 (poetry) | 33.26*** | 2 | 27.79*** | 2 |

*Significant at 10 per cent level only.
**Significant at 5 per cent level only.
***Significant at 0.1 per cent level.

methods are very different. In table 6.12 the goodness-of-fit of the expected frequencies obtained for the two estimates of the pair $(p, k)$ is given.

Thus, with the exception of *Portrait*, *Time*, and *Henry IV*, part 2, a reasonable fit is obtained by at least one of the methods of estimation.

In taking a second sample of 515 words from *Portrait*, we found 360 one-syllable words, 121 two-syllable words, 31 three-syllable words, 2 four-syllable words, and 1 five-syllable word. Thus the mean number of syllables is 1.3748 and the sample standard deviation is $s = 0.6341$, which is quite consistent with the sample in table 6.9. Using these values to estimate $p$ and $k$, we obtain

$$\hat{p} = 0.9321 \quad \text{and} \quad \hat{k} = 5.1451,$$

which yields $\chi^2 = 1.79$ in the goodness-of-fit test (with one degree of freedom) which is not significant at even the 10 per cent level. If we use these values of $p$ and $k$ to compute the expected frequencies for a sample of size 539 and compare them to the observed values in table 6.9, we obtain $\chi^2 = 3.36$ with three degrees of freedom (since the parameters were not computed from the data of table 6.9). $P(\chi_3^2 \geq 3.4) = 0.33397$, so the approximation is relatively good in this case.

In two more samples of 500 words each from *Time*, we obtained the frequencies given in table 6.13. From sample 1 in table 6.13, using the mean and standard deviation, we obtain estimates of $p$ and $k$ as follows:

$$\hat{p} = 0.6733 \quad \text{and} \quad \hat{k} = 1.5584.$$

If the expected frequencies using the negative binomial distribution are computed and a goodness-of-fit test applied (to sample 1), we obtain $\chi^2 = 40.09$ with two degrees of freedom. If the first two sample frequencies are used to estimate $\hat{p}$ and $\hat{k}$, the fit obtained is worse than the above, which is itself highly

*I*

TABLE 6.13

Two more samples from *Time*

| Syllables per word | Sample 1 | Sample 2 |
|---|---|---|
| 1 | 294 | 325 |
| 2 | 89 | 97 |
| 3 | 69 | 59 |
| 4 | 41 | 11 |
| 5 | 7 | 5 |
| 6 | 0 | 2 |
| 7 | 0 | 0 |
| 8 | 0 | 1 |
| TOTAL | 500 | 500 |
| $\bar{x}$ | 1.7560 | 1.5700 |
| $s$ | 1.0596 | 0.9481 |

significant. For sample 2 in table 6.13, we obtain estimates $\hat{p} = 0.6341$ and $\hat{k} = 0.9879$, using the sample mean and standard deviation, and a goodness-of-fit $\chi^2$ of 11.24 with two degrees of freedom. Since $P(\chi_2{}^2 \geq 11) = 0.00409$, the departure of the observed from the expected negative binomial frequencies is significant. The fit for the other estimates of $p$ and $k$ is again somewhat worse. One reason for these poor fits to the negative binomial frequencies of the frequencies observed for *Time* might be that many authors may have been involved, so that there is heterogeneity within the samples. If we test the three *Time* samples, i.e., that of table 6.9 and the two in table 6.13, for homogeneity, we find that $\chi^2 = 36.5$ with eight degrees of freedom. Since $P(\chi_8{}^2 \geq 36) = 0.00002$, *Time* appears to exhibit a very significant amount of heterogeneity among samples, which, as we have said, may be due to the multiplicity of authorship and the variability in topics.

Now let us consider the problem of *The Lady of the Lake*. In a second sample of 1000 words composed of 10 groups of 100 consecutive words selected at random within the poem, the relative frequencies obtained were 780, 191, 28, and 1 for one-, two-, three-, and four-syllable words respectively. The sample mean and standard deviation are 1.2503 and 0.4990 respectively, which resemble

TABLE 6.14

Relative frequencies in two samples from *The Lady of the Lake*

| Syllables per word | 1st sample | 2nd sample |
|---|---|---|
| 1 | 0.764 | 0.780 |
| 2 | 0.214 | 0.191 |
| 3 | 0.022 | 0.028 |
| 4 | 0.000 | 0.001 |

TABLE 6.15
Fitting data from *The Lady of the Lake*

| Syllables per word | Observed frequency | Expected frequency | |
|---|---|---|---|
| | | Translated Poisson | Negative binomial |
| 1 | 382 | 389.28 | 390.05 |
| 2 | 107 | 97.44 | 95.49 |
| $\geqq 3$ | 11 | 13.28 | 14.46 |
| TOTAL | 500 | 500.00 | 500.00 |
| DF | | 2 | 2 |
| $\chi^2$ | | 1.47 | 2.38 |

closely those obtained in the first *Lady of the Lake* sample in table 6.9. If we take the relative frequencies in each case, we obtain table 6.14.

Because of the small number of categories available in the syllables per word samples, the estimation of parameters $p$ and $k$ from the same set of samples as that from which the observed frequencies are obtained would render the degrees of freedom ridiculously small. Therefore, it is to our advantage to take two samples and obtain the estimates from one and the relative frequencies from the other. For example, we could use the first (500-word) sample of *The Lady of the Lake* to estimate the parameters and the second (1000-word) sample to obtain the sample frequencies or vice versa. In the present case, we must do the latter because it is impossible to obtain the negative binomial parameters for the first sample. Using the second sample, we obtain estimates $\hat{\lambda} = 1.2503$, and $\hat{p} = 0.9796$ and $\hat{k} = 12.0007$ (using the first two frequencies and the mean). The results are displayed in table 6.15. Note, in this case, that the translated Poisson distribution appears to yield the better fit.

For the poetry in *Henry IV*, part 2, it is clear, because we are dealing with the entire population, that it cannot be classified as either a translated Poisson or a translated negative binomial population.

TABLE 6.16
Data from *Julius Caesar*

| Syllables per word | Act I | Act II | Act III | Act IV | Act V |
|---|---|---|---|---|---|
| 1 | 71 | 80 | 78 | 83 | 76 |
| 2 | 24 | 17 | 14 | 13 | 16 |
| 3 | 5 | 3 | 8 | 3 | 4 |
| 4 | | | | 1 | 4 |
| TOTAL | 100 | 100 | 100 | 100 | 100 |
| $s$ | 0.5724 | 0.4894 | 0.6113 | 0.5427 | 0.7456 |
| $\bar{x}$ | 1.3400 | 1.2300 | 1.3000 | 1.2200 | 1.3600 |

In general, however, our data indicate that a reasonable case can be made for the hypothesis that the frequencies of syllables per word follow the negative binomial distribution. The exceptions, *Time*, *The Lady of the Lake*, and the poetry in *Henry IV*, part 2, might be explained by heterogeneity in the first case and some special features of the poetical forms in the latter cases.

Before we can be completely satisfied with the negative binomial hypothesis, it will be necessary to develop a model which explains *why* the syllable counts should follow the (translated) negative binomial distribution and what the significance of the parameters $p$ and $k$ might be.

### 6.3.2 *Homogeneity within works and authors*

Now let us turn our attention to the problem of consistency in the syllables per word distributions within works. Eight works are under consideration: *Julius Caesar* and *As You Like It* by Shakespeare, *The Cocktail Party* and *Four Quartets* by T.S. Eliot, *Victory* and *Lord Jim* by Joseph Conrad, and *The Solid Mandala* and *Riders in the Chariot* by Patrick White.

In five samples of 100 consecutive words, one chosen from each act of *Julius Caesar*, we obtained the results in table 6.16. The $\chi^2$-test for homogeneity yields $\chi^2 = 9.29$ with eight degrees of freedom, where $P(\chi_8^2 \geq 9.2) = 0.32571$. Thus we lack evidence for heterogeneity. Similar results are obtained from *As You Like It* (see table 6.17), where $\chi^2 = 8.90$ with again eight degrees of freedom. Thus both Shakespearian items exhibit a fair degree of homogeneity.

Combining the Shakespeare data and testing for homogeneity among the 10 samples, we obtain $\chi^2 = 18.38$ with 18 degrees of freedom. In this case $P(\chi_{18}^2 \geq 18) = 0.45565$. Thus homogeneity seems to exist despite the fact that the two plays belong to vastly different genres.

On the face of it, these results appear to contradict the impression which might be obtained if we test the results obtained in Elderton (1949) for *Henry IV*, part 2 (see table 6.10). In testing for homogeneity, we obtain $\chi^2 = 20.63$ with only three degrees of freedom, which is significant at the 0.00005 level. This high

TABLE 6.17

Data from *As You Like It*

| Syllables per word | Act I | Act II | Act III | Act IV | Act V |
|---|---|---|---|---|---|
| 1 | 77 | 76 | 81 | 84 | 75 |
| 2 | 19 | 19 | 15 | 8 | 20 |
| 3 | 4 | 4 | 2 | 8 | 5 |
| 4 |  | 1 | 2 |  |  |
| TOTAL | 100 | 100 | 100 | 100 | 100 |
| $s$ | 0.5291 | 0.5946 | 0.5925 | 0.5881 | 0.5596 |
| $\bar{x}$ | 1.2700 | 1.3000 | 1.2500 | 1.2400 | 1.3000 |

TABLE 6.18

Test for homogeneity among Shakespeare works

| Syllables per word | *Henry* IV, part 2 | *J.C.* | *A.Y.L.I.* | Expected relative frequency |
|---|---|---|---|---|
| 1 | 20,041 | 388 | 393 | 0.7923 |
| 2 | 4,095 | 84 | 81 | 0.1621 |
| 3 | 906 | 23 | 23 | 0.0362 |
| 4 | 238 | 5 | 3 | 0.0094 |
| TOTAL | 25,280 | 500 | 500 | 1.0000 |

$\chi^2 = 3.69$, DF $= 6$, $P(\chi_6^2 \geqq 3.8) = 0.70372$.

value of $\chi^2$ could have been obtained because of the large sample size in relation to the number of degrees of freedom. One way to test this possibility would be to take a number of 100-word samples from *Henry IV*, part 2, and test them for homogeneity. Oddly enough if we combine the *Henry IV* samples and test this sample with the combined *Julius Caesar* samples and the combined *As You Like It* samples, we obtain the results of table 6.18, which indicate a high degree of homogeneity among the three works.

In 10 random samples of 100 consecutive words from *The Cocktail Party*, where five samples were taken from act I and three and two respectively from acts II and III, we obtained the results in table 6.19. Testing the acts separately for homogeneity, we obtain: for act I, $\chi^2 = 15.20$ with eight degrees of freedom; for act II, $\chi^2 = 14.25$ with 12 degrees of freedom; for act III, $\chi^2 = 1.16$ with three degrees of freedom. The act I $\chi^2$ is almost significant at the 0.05 level, the act II $\chi^2$ could be obtained in more than one time in four, and the act III $\chi^2$ shows a high degree of homogeneity. If we test all the samples together for homogeneity, $\chi^2 = 32.34$ with eight degrees of freedom, whence $P(\chi_{18}^2 \geqq 32) = 0.02199$. Thus the play as a whole seems to indicate a significant lack of homogeneity.

TABLE 6.19

Syllable count in *The Cocktail Party*

| Syllables per word | Act I | | | | | Act II | | | Act III | |
|---|---|---|---|---|---|---|---|---|---|---|
| | 1 | 2 | 3 | 4 | 5 | 6 | 7 | 8 | 9 | 10 |
| 1 | 81 | 79 | 75 | 80 | 75 | 71 | 87 | 68 | 78 | 73 |
| 2 | 18 | 20 | 21 | 15 | 15 | 20 | 9 | 21 | 15 | 16 |
| 3 | 1 | 1 | 1 | 3 | 5 | 3 | 1 | 7 | 4 | 7 |
| 4 | | | 3 | 2 | 4 | 3 | 3 | 4 | 3 | 3 |
| 5 | | | | | 1 | 3 | | | | 1 |
| TOTAL | 100 | 100 | 100 | 100 | 100 | 100 | 100 | 100 | 100 | 100 |
| $s$ | 0.4264 | 0.4399 | 0.6495 | 0.6172 | 0.8420 | 0.9261 | 0.6030 | 0.7972 | 0.6946 | 0.8319 |
| $\bar{x}$ | 1.2000 | 1.2200 | 1.3200 | 1.2700 | 1.4100 | 1.4700 | 1.2000 | 1.4700 | 1.3200 | 1.4300 |

TABLE 6.20
Syllable counts from *Four Quartets*

| Syllables per word | 1st fifth | 2nd fifth | 3rd fifth | 4th fifth | 5th fifth |
|---|---|---|---|---|---|
| 1 | 72 | 75 | 71 | 73 | 76 |
| 2 | 18 | 19 | 21 | 23 | 20 |
| 3 | 5 | 6 | 6 | 3 | 4 |
| 4 | 3 | | 2 | 1 | |
| 5 | 2 | | | | |
| TOTAL | 100 | 100 | 100 | 100 | 100 |
| $s$ | 0.8805 | 0.5808 | 0.6948 | 0.5840 | 0.5333 |
| $\bar{x}$ | 1.4500 | 1.3100 | 1.3900 | 1.3200 | 1.2800 |

Now let us look at *Four Quartets*. We divided the work into fifths and took one sample of 100 consecutive words at random from each fifth to obtain the results of table 6.20. The homogeneity test yields $\chi^2 = 5.22$ with eight degrees of freedom, indicating a reasonable degree of homogeneity.

Turning to Conrad, we find the story is somewhat more complicated. To test for homogeneity in *Victory*, we took 10 samples of 100 consecutive words at random throughout the whole text with the results indicated in table 6.21. This time the homogeneity test yielded $\chi^2 = 9.41$ with 18 degrees of freedom. Since $P(\chi_{18}^2 \geq 9.4) = 0.94974$, we can take this as evidence for homogeneity.[4]

A similarly taken sample of 10 passages from *Lord Jim* (table 6.22) yields $\chi^2 = 19.38$ with 18 degrees of freedom. Since $P(\chi_{18}^2 \geq 19.5) = 0.36166$ in this instance, we have no evidence against homogeneity here either.

In testing all 20 samples from *Victory* and *Lord Jim*, taken together, for homogeneity, we obtain $\chi^2 = 53.87$ with 38 degrees of freedom, which is significant at the 5 per cent level. This is possibly due to a change in Conrad's style between the publication of *Lord Jim* (1899) and the publication of *Victory* (1914).

Finally we consider the Patrick White works in regard to homogeneity. In two samples of 1000 and 2000 consecutive words taken at random from *The Solid Mandala*, we obtain the results in columns 1 and 2 of table 6.23. In testing for homogeneity, we obtain $\chi^2 = 3.86$ with four degrees of freedom. Thus $P(\chi_4^2 \geq 3.8) = 0.43375$. In collecting one random sample of 500 consecutive words from each of five equal parts of *The Solid Mandala*, we obtain columns 3–7 of table 6.23. A homogeneity test of these samples, together with the two mentioned above, yields $\chi^2 = 13.96$ with 18 degrees of freedom. Thus with $P(\chi_{18}^2$

4  In a sample composed of the first 10 consecutive passages of 100 consecutive words from *Victory*, we tested for homogeneity, obtaining $\chi^2 = 49.47$ with 27 degrees of freedom. Since $P(\chi_{27}^2 \geq 48) = 0.00768$, there is a good case for heterogeneity here. However, this sample might not be representative of the style of the work as a whole because, for example, parts of it might have been worked over by the author more than the material in the rest of the book.

TABLE 6.21
A random sample of 10 passages from *Victory*

| Syllables per word | 1 | 2 | 3 | 4 | 5 | 6 | 7 | 8 | 9 | 10 | Expected value |
|---|---|---|---|---|---|---|---|---|---|---|---|
| 1 | 73 | 78 | 74 | 73 | 76 | 75 | 68 | 72 | 71 | 64 | 72.3 |
| 2 | 19 | 15 | 16 | 17 | 15 | 17 | 21 | 20 | 17 | 25 | 18.2 |
| 3 | 5 | 6 | 10 | 4 | 7 | 5 | 9 | 7 | 11 | 7 | 7.1 ⎫ |
| 4 | 2 | 0 |  | 3 | 2 | 2 | 0 | 1 | 1 | 3 | 1.4 ⎬ 9.5 |
| 5 | 1 | 1 |  | 3 | 2 | 1 | 2 | 1 | 1 | 1 | 1.0 ⎭ |
| TOTAL | 100 | 100 | 100 | 100 | 100 | 100 | 100 | 100 | 100 | 100 | |
| s | 0.7669 | 0.6771 | 0.6594 | 0.9366 | 0.7017 | 0.7608 | 0.8221 | 0.6614 | 0.7272 | 0.8346 | |
| x̄ | 1.3900 | 1.3100 | 1.3600 | 1.4600 | 1.3500 | 1.3700 | 1.4700 | 1.3700 | 1.4200 | 1.5200 | |

$\chi^2 = 9.4129$, DF $= 18$.

TABLE 6.22
A random sample of 10 passages from *Lord Jim*

| Syllables per word | 1 | 2 | 3 | 4 | 5 | 6 | 7 | 8 | 9 | 10 | Expected value |
|---|---|---|---|---|---|---|---|---|---|---|---|
| 1 | 72 | 77 | 79 | 78 | 79 | 78 | 73 | 72 | 72 | 74 | 76.4 |
| 2 | 16 | 13 | 10 | 15 | 9 | 15 | 20 | 22 | 21 | 20 | 16.1 |
| 3 | 9 | 4 | 9 | 5 | 9 | 7 | 4 | 5 | 6 | 6 | 6.4 ⎫ |
| 4 | 2 | 3 | 1 | 2 | 1 |  | 2 | 1 | 0 |  | 1.2 ⎬ 8.5 |
| 5 | 1 | 3 | 1 |  | 1 |  | 1 |  | 1 |  | 0.9 ⎭ |
| 6 |  |  |  |  | 1 |  |  |  |  |  | |
| TOTAL | 100 | 100 | 100 | 100 | 100 | 100 | 100 | 100 | 100 | 100 | |
| s | 0.8204 | 0.9340 | 0.7703 | 0.6620 | 0.8978 | 0.5911 | 0.7491 | 0.6256 | 0.6913 | 0.5840 | |
| x̄ | 1.4400 | 1.4200 | 1.3500 | 1.3100 | 1.3900 | 1.2900 | 1.3800 | 1.3500 | 1.3700 | 1.3200 | |

$\chi^2 = 19.3803$, $n = 30$, DF $= 18$.

TABLE 6.23

Data from *The Solid Mandala*

| Syllables per word | 1 | 2 | 3 | 4 | 5 | 6 | 7 |
|---|---|---|---|---|---|---|---|
| 1 | 710 | 1371 | 346 | 356 | 330 | 349 | 341 |
| 2 | 224 | 467 | 105 | 101 | 124 | 118 | 120 |
| 3 | 47 | 111 | 38 | 31 | 32 | 24 | 29 |
| 4 | 12 | 35 | 8 | 10 | 12 | 8 | 3 |
| 5 | 7 | 16 | 3 | 2 | 2 | 1 | 7 |
| TOTAL | 1000 | 2000 | 500 | 500 | 500 | 500 | 500 |
| $s$ | 0.6975 | 0.7566 | 0.7528 | 0.7304 | 0.7549 | 0.6741 | 0.7525 |
| $\bar{x}$ | 1.3820 | 1.4305 | 1.4340 | 1.4020 | 1.4640 | 1.3880 | 1.4300 |

$\geqq 13.5) = 0.76106$, we find no evidence against homogeneity here either. In *Riders in the Chariot*, we obtain similar results (consec. 1 and 2 in table 6.24) with a homogeneity $\chi^2 = 2.49$ with three degrees of freedom. Thus again no case can be made for heterogeneity.

Considering the two works together, we need not perform a full-scale homogeneity test involving the seven *S.M.* samples and the two *Riders* samples. We might first try to see if the *Riders* samples agree with the relative frequencies obtained from the *S.M.* homogeneity test performed above. These are 0.6915, 0.2289, 0.0567, and 0.0229 for words of one, two, three, and four or more syllables respectively. If the expected frequencies in the two *Riders* samples are obtained by multiplying these relative frequencies by 500 and a goodness-of-fit $\chi^2$ computed, we obtain $\chi^2 = 4.53$ with $(2 \times 4) - 1 = 7$ degrees of freedom (because the relative frequencies were not computed from the data this time). Since $P(\chi_7^2 \geqq 4.6) = 0.70864$, we see that the *Riders* samples conform very well to the *S.M.* frequency pattern, and so there is evidence for homogeneity.

Thus, except for *The Cocktail Party*, the works exhibit a fair degree of

TABLE 6.24

Data from *Riders in the Chariot*

| Syllables per word | Consec. 1 | Consec. 2 | Alt. 1 | Alt. 2 | Expected frequency |
|---|---|---|---|---|---|
| 1 | 347 | 354 | 343 | 363 | 351.75 |
| 2 | 111 | 102 | 105 | 88 | 101.50 |
| 3 | 35 | 31 | 35 | 35 | 34.00 |
| 4 | 6 | 12 | 12 | 12⎫ | 12.75 |
| 5 | 1 | 1 | 5 | 2⎭ | |
| TOTAL | 500 | 500 | 500 | 500 | |
| $s$ | 0.6915 | 0.7312 | 0.8086 | 0.7576 | |
| $\bar{x}$ | 1.4060 | 1.4080 | 1.4620 | 1.4040 | |

homogeneity when considering variations of syllable counts within the works. If the works tested are any indication, then the works of Shakespeare and Patrick White exhibit a high degree of homogeneity in regard to syllable count, while the homogeneity among the works of Conrad and Eliot, in this regard, is not so clear cut.

### 6.3.3 *A comparison of sampling methods*

All our samples of syllables per word have been samples of consecutive words. Can these be considered as random samples? More specifically, can we consider the number of syllables in a pair of consecutive words as independent of one another? One of the crudest ways of checking this is to sample alternate words only and then see if any difference is observed between these samples and consecutive samples. This was done for the four *Riders in the Chariot* samples in table 6.24. The usual homogeneity test yields $\chi^2 = 7.95$ with nine degrees of freedom, which could have arisen by chance in more than 53 cases in 100, thus offering no evidence of dependence.

For *Lord Jim* we obtained two samples at random of 500 consecutive words, four samples at random of 500 words taken alternately, and one sample of 500 words taken totally at random within the text. These are displayed on table 6.25. The homogeneity-$\chi^2$ equals 22.31, which could have been obtained 22 times out of 100 under the null hypothesis of homogeneity. Thus there appears to be little evidence here against the hypothesis of independence.

There is another way of checking the randomness of consecutive samples – the runs test discussed in section 5.5.

We have no sample of runs in *Riders in the Chariot*, but we do have such a sample of 1000 consecutive words from *The Solid Mandala*. In that sample there were $r = 205$ runs of one-syllable words, $n_0 = 714$ one-syllable words, and $n_1 = 286$ polysyllabic words.

In section 5.5 we observed that the number of runs of one-syllable words is

TABLE 6.25
Data from *Lord Jim*

| Syllables per word | Consec. 1 | Consec. 2 | Alt. 1 | Alt. 2 | Alt. 3 | Alt. 4 | Random | ROW TOTAL | Expected frequency |
|---|---|---|---|---|---|---|---|---|---|
| 1 | 366 | 383 | 347 | 349 | 375 | 383 | 364 | 2567 | 366.70 |
| 2 | 90 | 83 | 92 | 93 | 78 | 77 | 92 | 605 | 86.45 |
| 3 | 28 | 22 | 42 | 44 | 32 | 24 | 30 | 222 | 31.40 |
| 4 | 15 | 8 | 14 | 10 | 11 | 12 | 12 | 82⎫ | 15.15 |
| ≧ 5 | 1 | 4 | 5 | 4 | 4 | 4 | 2 | 24⎭ | |
| TOTAL | 500 | 500 | 500 | 500 | 500 | 500 | 500 | 3500 | |

$\chi^2 = 22.31$, DF $= 18$.

hypergeometrically distributed in the sense that the probability of $r$ runs is given by the formula

$$P(r) = h(r; n_0, n_1 + 1, n_0 + n_1).$$

Using the normal approximation to the hypergeometric distribution, the estimates of $\mu$ and $\sigma$ are given by (2) and (3) in section 5.5:

$$\hat{\mu} = 207.77, \qquad \hat{\sigma} = 6.51.$$

We find in this case that the probability of a greater deviation from the mean of 207.77 than in the present case is given by the expression

$$2\left\{1 - N\left(\left|\frac{207.77 - 205}{6.51}\right|; 0, 1\right)\right\} = 2\{1 - N(0.43; 0, 1)\} = 0.67$$

where $N(x; 0, 1)$ stands for the area to the left of $x$ under the unit normal curve. Thus this sample offers no evidence against randomness.

Turning to *Lord Jim*, we find that in a combined random sample of 10 passages of 100 consecutive words each, there were $r = 201$ runs of one-syllable words, $n_0 = 757$ one-syllable words, and $n_1 = 243$ polysyllabic words. The probability of a greater departure from the mean of 184.708 under the null hypothesis of randomness is

$$2\left\{1 - N\left(\frac{201 - 184.708}{5.8281}; 0, 1\right)\right\} = 2\{1 - N(2.7954; 0, 1)\} = 0.0051,$$

which is clearly significant.

The departure obtained above (in *L.J.*) seems to contradict the results obtained previously regarding the homogeneity of consecutive and alternate samples. One possible explanation is that independence obtains to a large extent between consecutive words, but there may be gross changes in the over-all distribution of the syllable count for the individual words as the style changes from one kind of writing (say narrative) to another (say conversation). Since the null hypothesis of randomness includes the hypothesis that the distribution of syllable counts is constant from word to word throughout the sample, this lack of uniformity may be the factor which accounts for the lack of randomness. A more thorough investigation would be necessary in order to resolve this question. It may turn out that some authors' styles are more homogeneous than others.

### 6.3.4 *Syllable counts for nouns*

We have counted the number of syllables in 1000 consecutive nouns (consecutive in the sense that no intervening nouns occur in the text) in two works, *Lord Jim* and *Riders in the Chariot*. The results appear in table 6.26. Here we have computed the expected frequencies for both samples under the two null hypotheses

TABLE 6.26

Data from a count of 1000 consecutive nouns in *Lord Jim* and in *Riders in the Chariot*

| Syllables per word | | L.J. | | | Riders | |
|---|---|---|---|---|---|---|
| | Obs. | Trans. Poisson exp. | Neg. binom. exp. | Obs. | Trans. Poisson exp. | Neg. binom. exp. |
| 1 | 508 | 457.49 | 498.41 | 532 | 522.05 | 540.38 |
| 2 | 303 | 357.76 | 310.46 | 337 | 339.33 | 315.25 |
| 3 | 102 | 139.88 | 128.24 | 93 | 110.28 | 108.11 |
| 4 | 73 | 36.46 | 44.00 | 27 | 23.90 | 28.41 |
| $\geqq 5$ | 14 | 8.41 | 18.89 | 11 | 4.45 | 7.85 |
| TOTAL | 1000 | | | 1000 | | |
| $\bar{x}$ | 1.7820 | | | 1.6500 | | |
| $s$ | 0.9907 | | | 0.8510 | | |

(i) that the sample comes from a translated Poisson population and (ii) that the sample comes from a translated negative binomial population. The reader by now should be able to see by simple observation that the departures of the observed frequencies from the translated Poisson expected frequencies for *Lord Jim* are significant. In computing the $\chi^2$ for the negative binomial expected frequencies, we obtain $\chi^2 = 26.11$ with only two degrees of freedom. Thus neither distribution fits the *Lord Jim* data. A similar story obtains for the *Riders* data. For the translated Poisson frequencies $\chi^2 = 12.96$ with three degrees of freedom, and for the negative binomial frequencies $\chi^2 = 5.08$ with two degrees of freedom. The translated Poisson $\chi^2$ is significant at the 0.01 level, but the negative binomial $\chi^2$ is significant only at the 0.08 level. There appears to be a slight hope that further investigation might yield evidence for a translated negative binomial fit, at least in the case of *Riders*.

Even without statistics it seems clear that the mean number of syllables for nouns in English is higher than that for the language as a whole. This can be seen by comparing the results of tables 6.9 and 6.26. There is one curious exception – the set of samples from *Time* given in tables 6.9 and 6.13, where the average over-all syllables per word counts are comparable with the *Lord Jim* and *Riders* noun counts.

We have data on runs for *Lord Jim* only, and here the number of runs of one-syllable nouns is $r = 242$ in 1000 consecutive nouns with 508 one-syllable nouns in the sample. The probability of this much or more departure from the mean of 250.44 is

$$2\left\{1 - N\left(\left|\frac{242 - 250.44}{7.91}\right|; 0, 1\right)\right\} = 2\{1 - N(1.07; 0, 1)\} \approx 0.3,$$

which does not indicate a significant lack of randomness in the sample.

6.4

SOME LEXICOSTATISTICAL MODELS FOR TESTING AFFINITIES
BETWEEN LANGUAGES OF UNKNOWN ORIGIN

In previous chapters we have discussed the models of Ross and Chrétien for the determination of the degree of linguistic relationship between languages descended from the same mother language, using the coincidence of root morphemes having the same meaning. Much attention has already been given to this problem in the more recent past (Dyen et al. 1967; Dobson 1969; Sankoff 1970; and Brainerd 1970), but we shall not go into it here. Rather we shall consider a more fundamental problem. Suppose we have no information regarding the relationship between two languages $L_1$ and $L_2$. Can we develop a test which reliably predicts a relationship if one exists? This 'relationship' can be due to genetic facts, or extensive borrowing, or some other cause not fully understood.

The method involves finding the probability that a single pair of words (or morphemes; see example 2 below) selected at random from the two languages share certain characteristics, under the null hypothesis that the languages are statistically independent of one another. Then $N$ such pairs of words are selected at random from the two languages. This process can be explicated in terms of an urn model (section 2.4) as follows. There are two urns, $U_1$ containing balls representing words in $L_1$, and $U_2$ bearing an analogous relation to $L_2$. A single pair can be selected at random by dipping into both urns simultaneously and selecting a ball at random from each urn. If $N$ pairs are chosen in this way (with replacement), the number $n$ of pairs where the two words share one of the characteristics mentioned above is obtained and the probability of $n$ such coincidences can be computed. If this probability is small, then we reject the null hypothesis and search for causes for the relationship.

We start with relatively crude models which are progressively refined until we arrive at our final model toward the end of the section.

The problem of selecting pairs $(w_1, w_2)$ at random, under the constraint that $w_i \in L_i$ can be solved in such a way as to minimize errors of the first kind by selecting a list of meanings like, for example, one of Swadesh's lists (1955) and using the pairs of words, one from each language, corresponding to each of these meanings.

6.4.1 *Taking the phonologies of the languages into account*

Starting from the position taken by Cowan (1962), let us consider the following model of the situation. Suppose languages $L_1$ and $L_2$ have inventories of phonemes $A_1$ and $A_2$ respectively such that there are $a = |A_1 \cap A_2|$ shared phonemes, $b_1 = |A_1 - A_2|$ phonemes in $A_1$ that are not in $A_2$, and $b_2 = |A_2 - A_1|$ members of $A_2$ not in $A_1$. Suppose we are to make up at random two $n$-phoneme words, one from the vocabulary of each of the languages. What is the probability that the phonemes in $k$ of the $n$ phoneme-slots will agree? As a first approximation, let us assume that no specification of the form of the word is made, that within the phoneme inventory of each language all the phonemes are equally likely to be chosen to fill a given slot, and that each slot is filled independently of the others. These are of course drastic simplifications which we shall need to modify later on. In any case, choosing one particular phoneme-slot we note that the probability of an agreement in that slot is equal to the probability that the slot is filled in the $L_1$-word, say, by an element $\varphi$ in $A_1 \cap A_2$ times the probability that the corresponding slot in the $L_2$-word is also filled by $\varphi$. This is given by the expression

$$(1) \quad p = \frac{a}{a+b_1} \frac{1}{a+b_2}.$$

Assuming that the choice for each slot is independent of the other choices, the filling of the $n$ slots from the two vocabularies can be construed as a sequence of $n$ Bernoulli trials where success is interpreted as filling two corresponding slots with the same phoneme, and the probability of success is given by $p$ in expression (1). Thus the theory of the binomial distribution, developed in chapter 2, leads us to the conclusion that the probability of exactly $k$ agreements in the $n$ slots is given by the expression

$$b(k; n, p) = \binom{n}{k} p^k (1-p)^{n-k}.$$

Then the probability of $k$ or more agreements in an $n$-slot word is the sum of the probabilities of the $n-k+1$ mutually exclusive outcomes favourable to $k$ or more agreements, namely

$$(2) \quad \pi_{k,n} = \sum_{j=k}^{n} b(j; n, p).$$

Now for a random list of meanings, $m_1, m_2, ..., m_q$, let $w_{rs}$ be the word in $L_r$ associated with meaning $m_s$. If we assume that the first $n$ phonemes (for example) of $w_{1s}$ and $w_{2s}$ are the result of a random selection from $A_1$ and $A_2$ respectively, then the probability of $k$ or more agreements among these phonemes is $\pi_{k,n}$. The probability of $k$ or more agreements among the first $n$ phonemes in exactly $l$ of the $q$ word pairs $(w_{1s}, w_{2s})$ is

$$(3) \quad \mathscr{P}_{l,k,n} = \binom{q}{l} (\pi_{k,n})^l (1-\pi_{k,n})^{q-l}.$$

To see how this model might be employed, consider the following example.

EXAMPLE 1   In comparing two West New Guinea languages, Sentani (S) and Demta (D), Cowan (1962) employs Swadesh's non-cultural list of meanings. This list contains meanings which are meant to be sufficiently fundamental to life and existence as to be expressed in a given language by words that represent the genetic development of the language. However, for our purposes Swadesh's list can be used as a random choice to select pairs of words one from each of the two languages we are testing.[5]

Cowan found that in Swadesh's list there were 75 meanings available for comparison, of which, according to him, there were 12 possible agreements. He suggests, however, that only the following 'straightest similarities' be used:

| S | D | Meaning |
|---|---|---------|
| wɛ | wé | those |
| nibi | nip | path, road |
| namə | nam-guai | three |
| naumə | namu | warm |
| bo | pu | bone |
| bu | po | water |

He finds that S contains 17 phonemes (7 vowels and 10 consonants) and D 23 (6 vowels and 17 consonants), and in common they contain 15 phonemes (6 vowels and 9 consonants). Thus, assuming $e \sim \epsilon$, there are 4 instances out of 75 where (among the first two phonemes) both agree.

According to (1)

$$p = 15/(17 \times 23) = 0.0384,$$

and according to (2)

$$\pi_{2,2} = p^2 = 0.0015.$$

If $X$ is the number of words out of the 75 non-cultural words where there is an agreement on the first two phonemes, then

$$(4) \quad P(X \geq 4) = \sum_{l=4}^{75} \binom{75}{l} (0.0015)^l (1 - 0.0015)^{75-l}.$$

Since $p$ is small, the Poisson distribution with

$$\mu = np = 75(0.0015) = 0.11$$

yields a good approximation to expression (4):

$$P(X \geq 4) = 1 - e^{-0.11} \left( 1 + 0.11 + \frac{(0.11)^2}{2} + \frac{(0.11)^3}{6} \right) = 0.000006.$$

This probability is unrealistically small, as we shall soon see.

5   In using this list we are making the possibility of errors of the first kind much smaller than if we used a more comprehensive list of meanings. However, we are, to the extent that Swadesh's theory is correct, testing our null hypothesis against the alternative of a genetic relationship rather than the more general alternative of a relationship due to no special cause in particular. More extensive lists, if they existed, might serve our purposes better.

TABLE 6.27

| Person | Ajamaru | Galela |
|---|---|---|
| 1 sg. | $t\,(e)$ | $t\,(o)$ |
| 2 sg. | $n\,(e)$ | $n\,(o)$ |
| 3 sg. m | $j\,(e)$ | $w\,(o)$ |
| 3 sg. f. | $m\,(e)$ | $m\,(o)$ |
| 3 sg. non-hum. | — | $i$ |
| 1 pl. exl. | — | $m\,(i)$ |
| 1 pl. incl. | $n\,(o)$ | $p\,(o)$ |
| 2 pl. | $n\,(e),\,b\,(o)$ | $n\,(i)$ |
| 3 pl. | $n\,(e),\,m\,(e)$ | $i$ |

EXAMPLE 2  Another example from Cowan (1962) compares the pronoun subject prefixes of the verb in Ajamaru (A) and in Galela (G) – both Austronesian
languages – to obtain the data in table 6.27. We might ask what the probability
of 4 agreements of initial consonants in 9 possible comparisons under our null
hypothesis of no relationship between the languages is. Since we are only considering consonants, vowels may be ignored. There are 13 consonants in A and
20 in G, of which 13 are shared according to Cowan. Assuming that the empty
initial consonant must be added to both languages, there are 21 in G and 14 in A.
If we treat the missing forms of A as simply different from the corresponding
forms in G, then our calculations can proceed as before. From (1)

$$p = \frac{14}{14 \times 21} = \frac{1}{21} = \pi_{1,1}.$$

Thus the probability of 4 or more agreements out of 9 is

$$P(X \geqq 4) = 1 - b\left(0; 9, \frac{1}{21}\right) - b\left(1; 9, \frac{1}{21}\right) - b\left(2; 9, \frac{1}{21}\right) - b\left(3; 9, \frac{1}{21}\right)$$

$$= 0.0010.$$

This probability, under the null assumptions of no causal relationship between
the languages and equal probability of consonant occurrence, indicates that the
null hypothesis embodied in these assumptions must be rejected. There is still a
question, however. Is it the lack of causal relationship between the languages,
or the assumption of equal probability of occurrence of the consonants, or both
that must be rejected? We shall come back to this problem in a moment.

### 6.4.2 Taking diagram regularities into account

The model in example 1 can be refined slightly by taking diagram regularities
into account if we assume that the initial two phonemes of the $w_{is}$ are always of
the form $cv$ and entertain the possibility that $c$ may be empty. Then there are
7 vowels and 11 consonants in S and 6 vowels and 18 consonants in D, with 6

vowels and 10 consonants in common. The probability of an initial-consonant agreement in $w_{1s}$ and $w_{2s}$ is

$$p_c = \frac{10}{11 \times 18} = 0.0505,$$

and the probability of a second-vowel agreement is

$$p_v = \frac{6}{6 \times 7} = \frac{1}{7} = 0.1429.$$

Assuming that the vowel choice and consonant choice are independent, the probability of the pair's agreeing is

$$p_c p_v = 0.0072.$$

The probability of 4 or more such agreements in 75 meanings is

$$P(X \geq 4) = \sum_{j=4}^{75} \binom{75}{4} (0.0072)^j (1-0.0072)^{75-j} = \sum_{j=4}^{75} \frac{(0.54)^j e^{-0.54}}{j!}$$

$$= 1 - e^{-0.54} \left( 1 + 0.54 + \frac{(0.54)^2}{2} + \frac{(0.54)^3}{6} \right) = 0.0023.$$

Although still significant, this result should be a more realistic estimate of the probability than the earlier one. However, the assumptions of equal probability of occurrence of consonants on the one hand and vowels on the other are still incorporated in the null hypothesis. We shall see a little later how we might take the relative frequencies of the individual phonemes into consideration. First let us consider another problem.

EXAMPLE 3   Sometimes one finds striking examples of similar single-word agreements in languages with no apparent connection. Using the model we have developed, we can attach a probability to the chance event of such occurrences under the null hypothesis of complete independence of the two languages, and hence make some assessment of the oddity of the event. As an example consider the words *mirar* in Spanish and *miru* in Japanese, which have highly similar meanings. We can obtain an estimate of the probability of the correspondence *mir*-S/*mir*-J by chance if we consider the probability of such occurrences among strings of the form *cvc* under the above null hypothesis. The structures of Spanish and Japanese phonology are remarkably similar so there are 19 consonants and 5 vowels in Spanish and 17 consonants and 5 vowels in Japanese, with 13 common consonants and 5 common vowels. Then the probability of a consonant correspondence is

$$p_c = \frac{13}{19 \times 17} = 0.04$$

and that of a vowel correspondence is

$$p_v = \frac{5}{5 \times 5} = 0.20.$$

Thus the possibility of the *mir-* being selected from all the *cvc* forms in both languages for (roughly) the same meaning by chance is

$$p_c p_v p_c = (0.2)(0.04)^2 = 0.0003.$$

However, even if the relative frequencies of the various phonemes in clusters of the form *cvc* in both languages were taken into consideration, the resulting probability that *mir-* would occur in both languages with roughly the same meaning would probably still be significant. A word of caution must be given about such methods as these. They should only be used as contributing evidence for causal relationships between languages. If a number of such rare events occur independently, all pointing toward such a relationship, then the researcher can begin to take the relationship seriously. Remember that just because a man gets four aces in a poker hand once, we do not accuse him of cheating. If, however, he gets three such hands in a row, we may have some reason to be uneasy.

To put the probability obtained in example 3 in a more correct perspective, suppose that in Swadesh's list of 100 non-cultural meanings only this one correspondence was obtained. Assuming for the sake of argument that all the words corresponding to Swadesh's meanings both in Spanish and Japanese were of the form *cvc*..., then the probability of one or more correspondences for the first three phonemes is

$$P(X \geqq 1) = 1 - (1 - p_c p_v p_c)^{100} \approx 1 - e^{-100(p_c p_v p_c)}$$

$$= 1 - 0.97 = 0.03.$$

In this setting the probability is 100 times the original form. If we had used Swadesh's 200-word list with only the one agreement, *mir-*, then

$$P(X \geqq 1) = 1 - (1 - p_c p_v p_c)^{200} \approx 1 - e^{-200(p_c p_v p_c)} = 0.06,$$

which is not quite significant at the 5 per cent level. Of course, as the number of words increases, the probability of one or more correspondences decreases. Indeed, with a test list of 1000 words,

$$P(X \geqq 1) = 1 - e^{-1000(p_c p_v p_c)} = 1 - 0.74 = 0.26.$$

In this case, if only the *mir-* correspondence occurred, it could not be viewed as a rare event at all. It goes without saying, then, that the longer the test list the more reliable the results, provided of course that the meanings are chosen independently of one another.

### 6.4.3 *Taking phoneme relative frequencies into account*

Examples 1, 2, and 3 indicate some of the imperfections in the model proposed at the beginning of the section. Most readers will probably have entertained at least one of the following pair of objections to this model during their reading so far.[6]

(i) Different ways in which researchers defined the phonemes in the two languages to be compared may have biased the probabilities of agreements at the phoneme level. This can happen in at least two ways: first in the actual choice of phonemes in the individual languages and second in the phonemes in one language which are chosen to be 'equal' to particular phonemes in the other. For example, language $A$ may have phonemes /p/, /b/, and /p'/ (aspirated p) and language $B$ may have /p/ and /b/ alone, although in the list of meanings there are a number of correspondences where the language $B$ forms begin with /b/ and the language $A$ forms begin with /p/ or /p'/. Suppose that the two languages agree on all the other 15 phonemes which are not bilabial stops. To use

$$p_c = \frac{15}{17 \times 18}$$

as the fundamental consonant frequency and to fail to count as agreements those correspondences that begin with /p/ and /p'/ in one case and /b/ in the other might yield a false picture.

(ii) The relative frequency with which the various phonemes in each language take part in words has not been taken into consideration. These relative frequencies will certainly alter the probabilities to a large extent. Both Swadesh and Cowan have observed this and have attempted to allow for it by multiplying their probabilities by a constant factor. However, this procedure only increases the vagueness of their results.

In the following pages a somewhat more subtle model is proposed in order to answer the objections just raised.

Let the two languages $L_1$ and $L_2$ contain inventories of 'words' $W_1$ and $W_2$ respectively. The elements of $W_i$ ($i = 1, 2$) are then strings of phonemes taken from a (finite) set $A_i$ as before. The intersection $A_1 \cap A_2$ may or may not be void. In most practical cases $W_i$ will also be a finite set. Now assume there is a designated collection $S_i$ ($= 2^{W_i}$ when $W_i$ is finite) of events, in this case subsets of $W_i$, and a probability $P_i$ defined on $S_i$ such that $S_i$ and $P_i$ form a probability space (in the sense of chapter 2) for $i = 1, 2$. Then for any set $G_i \in S_i$ of words from $W_i$, the probability of selecting a word in $G_i$ is $P(G_i)$.

Intuitively, in a practical application, $P(G_i)$ might be the ratio of the number

---

6  These objections, as we shall see, can be taken into consideration by altering the mathematical model. Others, such as the existence within a language of more than one word corresponding to a single meaning, cannot be so easily dealt with.

of elements of $G_i$ to the number of elements of $W_i$, and $G_i$ might be taken to be a union of sets of the form

(4) $G_{\varphi_1...\varphi_n; i} = \{w \in W_i \mid \text{first } n \text{ phonemes of } w \text{ are } \varphi_1 ... \varphi_n \text{ in that order}\}$

where $\varphi_1, ..., \varphi_n$ all belong to $A_1 \cap A_2$. For example, under the null hypothesis of no causal connection between $L_1$ and $L_2$, if we choose a string $\varphi_1\varphi_2 ... \varphi_n$ of $n$ phonemes from $A_1 \cap A_2$, then the probability of the event that $\varphi_1, \varphi_2, ..., \varphi_n$ constitute the first $n$ phonemes of a word in $W_i$ is

(5) $p_i(\varphi_1 ... \varphi_n) \overset{d}{=} P_i(G_{\varphi_1...\varphi_n; i})$

and the probability that two words selected at random, one from $W_1$ and one from $W_2$, both begin with $\varphi_1\varphi_2 ... \varphi_n$ is $p_1(\varphi_1 ... \varphi_n)p_2(\varphi_1 ... \varphi_n)$. Then the probability that two words $w_1$ and $w_2$, selected at random from $W_1$ and $W_2$ respectively, contain $n$ or more phonemes each and agree on their first $n$ phonemes is given by

(6) $p_n = \sum_{\varphi_1\varphi_2 ... \varphi_n} p_1(\varphi_1\varphi_2 ... \varphi_n)p_2(\varphi_1\varphi_2 ... \varphi_n)$

where the summation is taken over all strings $\varphi_1\varphi_2 ... \varphi_n$ of $n$ phonemes from $A_1 \cap A_2$.

Now suppose $q$ pairs $(w_1, w_2)$ of words are selected at random with $w_i \in W_i$. Let $X$ be the number of pairs where both entries contain at least $n$ phonemes and the first $n$ of these agree. Since each of the $q$ pairs is selected at random and the word banks $W_1$ and $W_2$ are large, these $q$ selections can be taken to be $q$ Bernoulli trials, and so

$$P(X = x) = \binom{q}{x} p_n^x(1-p_n)^{q-x}.$$

The basic difficulty of this model lies in the description of the probability spaces determined by $S_i$ and $P_i$. Sometimes, however, only a partial description of these spaces is needed in order to yield reasonable results, as the following example indicates.

EXAMPLE 4 Dyen (1964) gives the words in Lithuanian and Russian corresponding to 196 meanings of the original 200 in Swadesh's non-cultural list. Using this list to generate a (we hope) random sample of words in each of the languages, we can then estimate the values of $P_i$ for the sets of the form $G_{\varphi_1; i}$ defined by (4) as relative frequencies of elements of $G_{\varphi_1; i}$ in the list. Then estimates for the probabilities of the occurrence of arbitrary sets of initial consonants in each language can be computed using the laws of probability. If these sets are judiciously chosen, we can avoid having subjective judgments affect our results.

In the first column of table 6.28, we have indicated the initial consonants chosen. Except for the set of sibilants and velars and E, the empty initial con-

TABLE 6.28

| Initial consonants | Lithuanian relative frequency* | Russian relative frequency | PRODUCT |
|---|---|---|---|
| p | 0.0748 | 0.0638 | 0.0048 |
| b | 0.0234 | 0.0319 | 0.0007 |
| t | 0.0748 | 0.0798 | 0.0060 |
| d | 0.0794 | 0.0798 | 0.0063 |
| m | 0.0888 | 0.0585 | 0.0052 |
| n | 0.0561 | 0.0426 | 0.0024 |
| v | 0.0654 | 0.0638 | 0.0042 |
| Sibilants | 0.2383 | 0.2181 | 0.0520 |
| Velars | 0.2430 | 0.1649 | 0.0401 |
| l | 0.0327 | 0.0266 | 0.0009 |
| r | 0.0234 | 0.0479 | 0.0011 |
| E | 0.0000 | 0.1011 | 0 |
| t$^s$ | 0.0000 | 0.0053 | 0 |
| y | 0.0000 | 0.0160 | 0 |
| SUM OF PRODUCTS | | | 0.1237 |

* These initial-consonant frequency estimates were computed from the entire list of 196 words used by Dyen.

sonant, all the sets contain only one phoneme. The sibilants and velars are grouped together in order to avoid the necessity of making subjective decisions, and E corresponds to the situation in which the word is vowel-initial. The second and third columns contain the initial-consonant-probability estimates for each language.

From table 6.28 we see that the relative frequency of the event that two words selected at random, one from the Lithuanian list and one from the Russian list, contain first consonants from the same set in the first column of the table is 0.1237. If we use this relative frequency as an estimate of the probability $p$ that the first consonants of two words agree (i.e., lie in the same set in the first column), when they are selected at random, one from Lithuanian and the other from Russian, and if we look at the first 50 entires in Dyen's list, we find that in 23 cases out of 50 the first consonants agree. The probability of 23 such correspondences is (assuming that each meaning in Dyen's list engenders a random selection from the word-bank of each language) is given by the binomial probability law. If $X$ is the number of trials (meanings) for which the first consonants agree, then

$$P(X \geqq 23) = \sum_{i=23}^{50} \binom{50}{i} (0.1237)^i (1-0.1237)^{50-i}.$$

Since $np = 50(0.1237) = 6.1850$ is greater than 5, the normal approximation can be used with $\mu = 6.1850$ and $\sigma = \sqrt{np(1-p)} = 2.3281$. Therefore

$$P(X \geq 23) = 1 - N(22.5; 6.1850, 2.3281) = 1 - N(7.0079; 0, 1)$$
$$< 1 - N(4; 0, 1) = 0.00003.$$

Thus the hypothesis that the two languages are unrelated must be rejected if our first-consonant estimates are at all reliable.

The procedure in example 4 avoids the difficulties implicit in objections (i) and (ii) given on p. 262. The reader can easily see how this method can be employed in more equivocal instances to assess language relatedness.

The problem of the introduction of subjective bias due to the simultaneous use of the data for initial-consonant-frequency estimates and for cognate judgments can be obviated, in part at least, by dividing the list into two parts, one of which is used for the frequency estimates and the other for the cognate judgments.

In example 4 we have used only the monogram frequencies. If longer lists are used, then reasonable estimates of digram and trigram probabilities can be obtained and used in the method, thus increasing its efficiency.

## 6.5
### FURTHER READING

Because this book is intended to be only an introduction to the fundamental notions of classical statistics and probability theory that have application in the study of language and style, I have of necessity had to overlook certain topics which my readers may wish to pursue. There are also some topics that I have avoided because I feel they are either not properly in the realm of statistics in the sense of this book or of historical interest only and thus of little interest to researchers in the future. Of course, there is always a chance that one might be wrong in the latter assumption. I have, therefore, chosen to mention at this point some of the subjects that fall into one or both of these categories so that readers may pursue them if they wish.

A topic of this kind, to which a large amount of literature has been devoted, is the set of functional relationships between rank and frequency studied under the general heading of Zipf's law. For example, if we arrange the words (types) in a literary work, say *Ulysses* by James Joyce, in order of their frequency in the text, then with $w_i$ equal to the $i$th most frequent word, we find that the data fit the following relation quite closely:

(1) $w_i = ai^b$

where $a$ and $b$ are constants depending on the work under discussion. Relation (1) also fits other data as well. Indeed, if the cities of over 5000 people in the United States in 1940 are ranked according to size, then when $w_i$ is equal to the size of the $i$th ranking city, they satisfy (1) to a close degree for a certain choice of the parameters $a$ and $b$. Similarly, if the characters in a play are ranked according to the number of words they speak, they also tend to follow (1).

A large amount of space in the linguistic literature has been devoted to this phenomenon, but the results have been rather meagre. For the origin of the discussion, see Zipf (1965), and for a survey of further developments, see Cherry (1957, pp. 100–8 and 209–12) and Booth (1967), and finally for a critique of the enterprise and of others like it, see Simon (1968).

Closely related to Zipf's law is what is called the Waring-Herdan formula by Charles Muller in an article reprinted in Doležel and Bailey (1969, pp. 42–56). This article provides an interesting hypothesis concerning the probability $p_n$ that a lexical item, selected at random from a text, appears $n$ times in that text. For some general remarks on Zipf's law and some remarks about a distribution related to that of $p_n$ mentioned above, see I.J. Good's article on the statistics of language in Meetham and Hudson (1969, pp. 567–81).

Another topic, of perhaps historical interest only, which is usually grouped under the statistics of language is information theory as it is applied to language. Here again the results are only distantly related to statistics in our sense, that is to statistics as a collection of methods for making decisions in the face of un-certainty. The interested reader can begin with I.J. Good's article mentioned in the previous paragraph. He should also see chapter 5 of Cherry (1957). For an application of the theory to poetics see the article by Kondratov in Doležel and Bailey (1969, pp. 113–21).

A subject which should assume much greater importance as statistical methods in the study of language develop is the assignment of authorship in works where the author is unknown. For some remarks on the history of the authorship problem in its various forms, see R.W. Bailey's article in Doležel and Bailey (1969, especially pp. 222–7). Bailey also provides an ample bibliography on the subject.

By far the most successful attempt at authorship assignment is that in Mosteller and Wallace (1964), where the authors assess the probability that one of two possible authors wrote those Federalist Papers that are of unknown authorship. In this work they compare two statistical methods, one a so-called Bayesian approach to the problem and the other a classical discriminant analysis based on the work of R.A. Fisher (1936). Their main thrust was in the direction of a Bayesian analysis, based on an extended application of Bayes's rule (see our section 2.3). A heated controversy exists concerning the use of Bayes's rule in statistics and its philosophical implications. The statisticians who take an interest in the foundations of the subject divide themselves into two camps, the Bayesians (or Subjectivists) and the Frequentists. In our development of the concept of probability (chapter 2), we took a decidedly frequentist approach. For an ex-position of the Bayesian point of view, see Good (1965).

Mosteller and Wallace (1964) is a full technical account of the problem and its solution which may make heavy reading for the unprepared non-statistician.

I suggest, therefore, that the reader approach the work gradually by first reading the semi-popular account of their work by I.S. Francis in Leed (1966, pp. 38–78).

Other authorship studies, of varying degrees of success, which may be of interest (and are technically less demanding) are Brinegar (1963), the article by A.Q. Morton and Michael Levison in Leed (1966, pp. 141–79), the article by Dreher and Young in Doležel and Bailey (1969, pp. 156–69), Ellegård (1962), Yule (1944), and Williams (1970). See also Brainerd (1971). For a survey of additional authorship studies see Wachal (1966) as well as the bibliographies mentioned in the Preface.

In order to compare the work of pairs of authors, as in the study by Mosteller and Wallace (1964), or of many authors, as in Ellegård (1962), or to test whether it is plausible that a single author wrote a certain body of texts, as in Brinegar (1963), it is necessary to obtain valid information about the styles of individual authors. There are a number of interesting studies of the styles of individual authors in Doležel and Bailey (1969), and a survey of the subject appears in Bailey's essay on pages 220–1 of the volume. However, by far the most complete work of this kind to date is Milic (1967b) on Jonathan Swift.

During the development of what might loosely be called quantitative stylistics, much literature has been devoted to the study of special indices of style such as Yule's $K$, the *Aktionsquotient* or Verb-Adjective Ratio, the *hapax logomena*, and many others. The most famous of these is Yule's $K$. In Yule (1944), the author puts forward a theoretical development of $K$ giving some of its properties. Later, in Herdan (1966, pp. 101–4), the properties of $K$ are developed further. To see how $K$ behaves in a particular study, see the essay by Paul E. Bennett in Doležel and Bailey (1969, pp. 29–41). For a complete survey of such indices and their application to the study of specific texts, see Carpenter (1968).

A number of potentially useful topics have been omitted because their explicit development required mathematics beyond the capabilities of the intended audience of this book. Among these are factor-analysis, discriminant analysis, and cluster analysis. Programs for these techniques are available in most computer installations so that they can easily be performed, even by researchers with a limited knowledge of their capabilities. However, when interpreting the results of such studies, it is best to consult an expert before drawing any firm conclusions. For a study using factor-analysis, see J.B. Carroll's article in Doležel and Bailey (1969, pp. 147–55). Mosteller and Wallace (1964) provide an example of a discriminant analysis, and cluster analysis is used in Brainerd (1972). Cluster analysis provides a very potent tool for theory-constructing, and readers interested in the subject should begin by consulting Goodman and Kruskal (1954 and 1959) and the references mentioned in Brainerd (1972).

Closely related to the above topics is multiple regression, which we briefly mentioned in chapter 1. For an interesting use of this technique, see Bernard O'Donnell's article in Leed (1966, pp. 107–15).

# References

Bailey, R.W., and Burton, Dolores M. 1968. *English Stylistics: A Bibliography* (Cambridge, Mass. and London: MIT Press)

Bailey, R.W., and Doležel, L. 1968. *An Annotated Bibliography of Statistical Stylistics* (Ann Arbor: Department of Slavic Languages and Literatures, University of Michigan)

Becker, S.W., and Carroll, J. 1963. Effects of high and low sentence contingency on learning and attitude. *Language and Speech*, 6, 46–56

Booth, A.D. 1967. A 'law' of occurrences for words of low frequency. *Information and Control*, 10, 386–93

Bradley, J.V. 1968. *Distribution-free Statistical Tests* (Englewood Cliffs, N.J.: Prentice-Hall)

Brainerd, Barron. 1970a. *Introduction to the Mathematics of Language Study* (New York: American Elsevier)

— 1970b. A stochastic process related to language change. *J. Applied Probability*, 7, 69–78

— 1971. The computer in statistical studies of William Shakespeare. Report to the Consultative Committee on Shakespeare and the Computer, World Shakespeare Congress, Vancouver

— 1972. An exploratory study of pronouns and articles as indices of genre in English. *Language and Style*, 5, 239–59

Brainerd, Barron, Clarke, D.A., Liebovitz, M.J., Ross, R.A., and Scroggie, G.A. 1967. *Topics in Mathematics*, 3 vols. (London: Frederick Warne)

Brandt Corstius, H. 1970. *Exercises in Computational Linguistics* (Amsterdam: Mathematisch Centrum)

Brinegar, C.S. 1963. Mark Twain and the Quintus Curtius Snodgrass letters: a statistical test of authorship. *J. Amer. Stat. Assoc.*, 58, 85–96

Burkhill, J.C. 1961. *The Lebesque Integral.* Cambridge Tracts in Mathematics and Mathematical Physics, no. 40, chapters I and II

Carpenter, G.J. 1968. The interrelationships among various statistical measures of written language and their relative importance in discriminating between the work of two authors. Unpublished MA thesis, York University, Toronto

Cassotta, Louis, Feldstein, Stanley, and Jaffe, Joseph. 1964. Markovian model of time patterns of speech. *Science*, 144, 884–6

Chakravarti, I.M., Laha, R.G., and Roy, J. 1967. *Handbook of Methods of Applied Statistics*, vol. I (New York: Wiley)

Chatman, Seymour (ed.). 1971. *Literary Style: A Symposium* (London: Oxford University Press)

Cherry, Colin, 1957. *On Human Communication* (New York: Wiley)

Chrétien, C.D. 1943. The quantitative method for determining linguistic relationships *University of California Publications in Linguistics*, 1, 11–19

— 1956. Word distributions in southeastern Papua. *Language*, 32, 88–109

Chrétien, C.D., and Kroeber, A.L. 1937. Quantitative classification of Indoeuropean languages. *Language*, 13, 83–104

Courant, R.C., and Robbins, H.R. 1941. *What is Mathematics?* (London: Oxford University Press)

Cowan, H.K.J. 1962. Statistical determination of linguistic relationship. *Studia Linguistica*, 16, 57–96

Cox, D.R. 1970. *Analysis of Binary Data* (London: Methuen)

Dobson, A.J. 1969. Lexicostatistical grouping. *Anthropological Linguistics*, 11, 216–21

Doležel, L., and Bailey, R.W. (eds.). 1969. *Statistics and Style* (New York: American Elsevier)

Dyen, Isidore. 1964. On the validity of comparative lexicostatistics. *Proc. IXth International Congress of Linguists* (The Hague: Mouton), pp. 238–52

Dyen, Isidore, James, A.T., and Cole, J.W.L. 1967. Language divergence and estimated word retention rate. *Language*, 43, 150–71

Elderton, W.P. 1949. A few statistics on the length of English words. *J. Roy. Stat. Soc.*, ser. A, 112, 436–45

Ellegård, Alvar. 1959. Statistical measurement of linguistic relationship. *Language*, 35, 131–56

— 1962. *A Statistical Method for Determining Authorship: The Junius Letters, 1769–1772*. Gothenburg Studies in English, no. 13 (Gothenburg)

Fisher, R.A. 1936. The use of multiple measurements in taxonomic problems. *Annals of Eugenics*, 7, pt. 2, 179–88

Fucks, W. 1955. *Mathematische Analyse von, Sprachelementen, Sprachstil und Sprachen* (Cologne: West Deutscher Verlag)

— 1956. Mathematical theory of word formation. In *Information Theory: Third London Symposium*, ed. Colin Cherry (London: Butterworth), pp. 154–70

Garvin, P.L. (ed.). 1964. *A Prague School Reader on Esthetics, Literary Structure and Style* (Washington, D.C.: Georgetown University Press)

Gleason, H.A., Jr. 1955. *An Introduction to Descriptive Linguistics* (New York: Holt, Rinehart, and Winston)

Good, I.J. 1965. *The Estimation of Probabilities: An Essay on Modern Bayesian Methods* (Cambridge, Mass.: MIT Press)

Goodman, L.A., and Kruskal, W.H. 1954. Measures of association for cross classifications. *J. Amer. Stat. Assoc.*, 49, 732–64

— 1959. Measures of association for cross classifications, II: Further discussion and references. *J. Amer. Stat. Assoc.*, 54, 123–63

Green, J.A. 1958. *Sequences and Series* (London: Routledge and Kegan Paul)

Greenberg, Joseph H. 1956. The measurement of linguistic diversity. *Language*, 32, 109–16

Guiraud, Pierre. 1959. *Problèmes et méthodes de la statistique linguistique* (Dordrecht: Reidel)

Hamilton, R. 1949. *The Telltale Article* (London: William Heinemann)

Hayden, Rebecca E. 1950. The relative frequency of phonemes in general-American English. *Word*, 6, 217–23

Herdan, Gustav. 1966. *The Advanced Theory of Language as Choice and Chance* (New York: Springer-Verlag)

Hilton, P.J. 1958. *Differential Calculus* (London: Routledge and Kegan Paul)

Hodge, J.L., and Lehmann, E.L. 1964. *Basic Concepts of Probability and Statistics* (San Francisco: Holden Day)

Houck, Charles L. 1967. A computerized statistical methodology for linguistic geography. *Folia Linguistica*, 1, 80–95

Hultzen, L.S., Allen, J.H.D., and Miron, M.S. 1964. *Tables of Transition Frequencies of English Phonemes* (Urbana: University of Illinois Press)

Kemeny, J.G., Snell, J.L., and Thompson, G.L. 1957. *Introduction to Finite Mathematics* (Englewood Cliffs, N.J.: Prentice-Hall)

Kendall, D.G., and Stuart, A. 1966. *The Advanced Theory of Statistics*, vol. 3 (London: Griffin and Co.)

Krámský, J. 1966. The frequency of occurrence of vowel phonemes in languages possessing vowel systems of identical structure. *Prague Studies in Mathematical Linguistics*, 1, 17–31

— 1967. The frequency of articles in relation to style in English. *Prague Studies in Mathematical Linguistics*, 2, 89–95

Kučera, H., and Nelson Francis, W. 1967. *Computational Analysis of Present-Day American English* (Providence: Brown University Press)

Labov, William. 1966. *The Social Stratification of English in New York City* (Washington, D.C.: Washington Center for Applied Linguistics)

Lancaster, H.O. 1969. *The Chi-Squared Distribution* (New York: Wiley)

Leed, Jacob (ed.). 1966. *The Computer and Literary Style: Introductory Essays and Studies* (Kent, Ohio: Kent State University Press)

Lieberman, G.J., and Owen, D.B. 1961. *Tables of Hypergeometric Probability Distribution* (Palo Alto: Stanford University Press)

Markov, A.A. 1913. Essai d'une recherche statistique sur le texte du roman 'Eugene Onegin.' *Bull. Acad. Imper. Sci. St. Petersburg*, ser. 6, 7, 153–69

Meetham, A.R., and Hudson, R.A. (eds.). 1969. *Encyclopaedia of Linguistics, Information and Control* (Oxford: Pergamon Press)

Milic, L.T. 1967a. *Style and Stylistics: An Analytic Bibliography* (New York and London: Free Press)

— 1967b. *A Quantitative Approach to the Style of Jonathan Swift* (The Hague: Mouton)

Miller, G.A. 1951. *Language and Communication* (New York: McGraw-Hill)

Moreau, René. 1964. *Une méthode de décomposition en syllabes par ordinateur d'un mot d'une langue romane*, pt. II. Étude n° 0001, IBM Paris

Mosteller, F., and Wallace, D.L. 1964. *Inference and Disputed Authorship: The Federalist* (Reading, Mass.: Addison-Wesley)

Muller, Charles. 1968. *Initiation à la statistique linguistique* (Paris: Larousse)

Patil, G.G., and Joshi, S.W. 1968. *A Dictionary and Bibliography of Discrete Distributions* (New York: Hafner)

Pearson, E.S., and Hartley, H.O. 1962. *Biometrika Tables for Statisticians*, vol. I (Cambridge: Cambridge University Press)

Rand Corporation. 1955. *A Million Random Digits* (Glencoe, Ill.: The Free Press)

Read, David W. 1949. A statistical approach to quantitative linguistic analysis. *Word*, 5, 235–47

Roberts, A. Hood. 1965. *A Statistical Linguistic Analysis of American English* (The Hague: Mouton)

Ross, A.S.C. 1950. Philological probability problems. *J. Roy. Stat. Soc*, ser. B, 12, 19–59 (including discussion)

Sankoff, David. 1970. On the rate of replacement of word-meaning relationships. *Language*, 46, 564–9

Scheffé, Henry. 1959. *The Analysis of Variance* (New York: Wiley)

Sebeok, T.A. (ed.). 1960. *Style in Language* (Cambridge, Mass.: MIT Press)

Simon, H.A., 1968. On judging the plausibility of theories. In *Logic, Methodology, and Philosophy of Science*, vol. III, eds. B. van Rootselaar and J.F. Staal (Amsterdam: North-Holland), pp. 439–59

Somers, H.H. 1962. *Analyse Statistique du Style*, 2 vols. (Louvain: Nauwalaerts)

Suppes, P. 1957. *Introduction to Logic* (New York: Van Nostrand)

Swadesh, M. 1955. Towards greater accuracy in lexico-statistic dating. *Int. J. Amer. Linguistics*, 21, 121–37

Wachal, R.S. 1966. Linguistic evidence, statistical inference and disputed authorship. Unpublished PHD thesis, University of Wisconsin

Wadsworth, G.P., and Bryan, J.G. 1960. *Introduction to Probability and Random Variables* (New York: McGraw-Hill)

Wilks, S.S. 1962. *Mathematical Statistics* (New York: Wiley)

Williams, C.B. 1970. *Style and Vocabulary: Numerical Studies* (London: Griffin)

Williams, E.J. 1959. *Regression Analysis* (New York: Wiley)

Yule, G. Udny. 1944. *The Statistical Study of Literary Style* (Cambridge: Cambridge University Press)

Zierer, E. 1964. Distribution of nouns in a German text and its corresponding Spanish translation. *Keiryo Kokugo Gaku*, 30, 38–40

Zipf, G.K. 1965. *The Psycho-biology of Language*, 2nd ed. (Cambridge, Mass.: MIT Press)

# Index

Lightning Source UK Ltd.
Milton Keynes UK
UKHW030614210722
406167UK00006B/634